Florian Kühnel

Stochastic Cosmological Inflation

Florian Kühnel

Stochastic Cosmological Inflation

A Replica Field-Theoretical Study

Südwestdeutscher Verlag für Hochschulschriften

Impressum / Imprint

Bibliografische Information der Deutschen Nationalbibliothek: Die Deutsche Nationalbibliothek verzeichnet diese Publikation in der Deutschen Nationalbibliografie; detaillierte bibliografische Daten sind im Internet über http://dnb.d-nb.de abrufbar.
Alle in diesem Buch genannten Marken und Produktnamen unterliegen warenzeichen-, marken- oder patentrechtlichem Schutz bzw. sind Warenzeichen oder eingetragene Warenzeichen der jeweiligen Inhaber. Die Wiedergabe von Marken, Produktnamen, Gebrauchsnamen, Handelsnamen, Warenbezeichnungen u.s.w. in diesem Werk berechtigt auch ohne besondere Kennzeichnung nicht zu der Annahme, dass solche Namen im Sinne der Warenzeichen- und Markenschutzgesetzgebung als frei zu betrachten wären und daher von jedermann benutzt werden dürften.

Bibliographic information published by the Deutsche Nationalbibliothek: The Deutsche Nationalbibliothek lists this publication in the Deutsche Nationalbibliografie; detailed bibliographic data are available in the Internet at http://dnb.d-nb.de.
Any brand names and product names mentioned in this book are subject to trademark, brand or patent protection and are trademarks or registered trademarks of their respective holders. The use of brand names, product names, common names, trade names, product descriptions etc. even without a particular marking in this works is in no way to be construed to mean that such names may be regarded as unrestricted in respect of trademark and brand protection legislation and could thus be used by anyone.

Coverbild / Cover image: www.ingimage.com

Verlag / Publisher:
Südwestdeutscher Verlag für Hochschulschriften
ist ein Imprint der / is a trademark of
AV Akademikerverlag GmbH & Co. KG
Heinrich-Böcking-Str. 6-8, 66121 Saarbrücken, Deutschland / Germany
Email: info@svh-verlag.de

Herstellung: siehe letzte Seite /
Printed at: see last page
ISBN: 978-3-8381-3250-1

Zugl. / Approved by: Bielefeld, Bielefeld University, Dissertation, 2009

Copyright © 2012 AV Akademikerverlag GmbH & Co. KG
Alle Rechte vorbehalten. / All rights reserved. Saarbrücken 2012

»Das Höchste, wozu der Mensch gelangen kann, ist das Erstaunen.«

JOHANN WOLFGANG VON GOETHE

Abstract

In this thesis we apply methods from statistical physics to stochastic inflation. Those methods, the replica field theory and the Gaussian variational methods, have to our knowledge never been applied before in this context, and allow us to compute the power spectrum of a scalar test field in the most general set-up. It provides a framework to perform calculations in regions of arbitrarily large quantum fluctuations and may also serve as a starting point to address the issue of back reaction.

We first give an introduction to cosmological inflation, cosmological perturbation theory and cosmic microwave background anisotropies. Then we explain the idea of stochastic inflation, including some detailed derivations, and give an overview over major progress in this field. This is followed by an introduction to replica field theory, presented in a way directly applicable to stochastic inflation. Our work continues with a detailed calculation of the power spectrum of a scalar test field in a Friedmann Universe. We elaborate on the effect of the quantum fluctuations on the spectrum and derive explicit expressions showing its dependence on time and other important parameters. The effect of self-interactions and possible effects on the cosmic microwave background are discussed. We conclude with a summary of our results and give an outlook.

One part of our major results has been published in Phys. Rev. D **78**, 103501 (2008), where for the first time we present a replica field-theoretical approach to stochastic inflation in which we find a manifestation of the phenomena of so-called dimensional reduction. It implies under certain conditions inevitable infra-red divergencies of correlation functions on large-scales. These conditions are examined in detail in Phys. Rev. D **79**, 44009 (2009), where we find that generically for a wide class of circumstances the divergencies are pushed exponentially fast well beyond observable scales. A subsequent publication is dedicated to the inclusion of self-interaction within our non-perturbative replica framework. For a quartic self-coupling we find a damping of the power spectrum on large scales — an issue which has recently attracted significant attention. Our findings are fully consistent with those in the literature and may provide an explanation of certain features in the cosmic microwave background, and might also help to resolve some long-standing infra-red problems in inflationary cosmology.

Abriss

In dieser Arbeit wenden wir Methoden aus der statistischen Physik auf die stochastische Inflation an. Diese, die Replika-Feldtheorie und die Gauß'sche Variationsmethode, sind unseres Wissens nach noch nie zuvor in diesem Kontext angewendet worden, und erlauben, das Leistungsspektrum eines skalaren Testfeldes in großer Allgemeinheit zu bestimmen. Diese Methoden stellen einen Rahmen zur Verfügung, in welchem beliebige Quantenfluktuationen behandelt werden können, und könnten ein Startpunkt sein, um Rückwirkungseffekte bezüglich der Geometrie der Raumzeit einzubeziehen.

Einführungen in kosmologische Inflation, kosmologische Störungstheorie und Anisotropien des kosmischen Mikrowellenhintergrundes leiten diese Arbeit ein. Danach erklären wir die Idee der stochastischen Inflation, wobei wir auch einige detaillierte Herleitungen präsentieren, und geben einen Überblick über wichtige Fortschritte in diesem Feld. Es folgt eine Einführung in die Methoden der Replika-Feldtheorie, und zwar in einer Weise, die direkt anwendbar ist auf die stochastische Inflation. Unsere Arbeit wird fortgeführt durch eine explizite Berechnung des Leistungsspektrums eines skalaren Testfeldes in einem Friedmann-Universum. Wir untersuchen den Einfluss von Quantenfluktuationen auf dieses Spektrum und leiten explizite Ausdrücke her, welche die Abhängigkeiten von der Zeit, sowie von anderen wichtigen Parametern deutlich machen. Der Effekt von Selbst-Wechselwirkungen und mögliche Auswirkungen auf den kosmischen Mikrowellenhintergrund werden diskutiert. Wir schließen mit einer Zusammenfassung und geben einen Ausblick.

Ein Teil unserer Hauptresultate wurde in Phys. Rev. D **78**, 103501 (2008) publiziert, worin wir zum ersten Mal einen Replika-Feldtheorie-Zugang für die stochastische Inflation präsentieren und das Auftreten des Phänomens der sogenannten dimensionellen Reduktion zeigen. Dieses impliziert, dass unter bestimmten Bedingungen Infrarot-Divergenzen in Korrelationsfunktionen auftreten können. Die genauen Bedingungen hierfür haben wir in der Veröffentlichung Phys. Rev. D **79**, 44009 (2009) untersucht, worin wir zeigen konnten, dass für eine große Klasse von Ausgangssituationen diejenigen Bereiche, in denen diese Divergenzen auftreten, exponentiell schnell aus den beobachtbaren Regionen verschoben werden. Eine weitere Publikation werden wir dem Studium von Selbst-Wechselwirkungen innerhalb des Replika-Feldtheorie-Rahmens widmen. Für eine quartische Selbst-Wechselwirkung finden wir eine Dämpfung auf großen Skalen – ein Aspekt der in jüngerer Zeit Aufmerksamkeit auf sich gezogen hat. Unsere Resultate sind konsistent mit denen aus der Literatur und könnten dazu beitragen, gewisse Fragestellungen zum kosmischen Mikrowellen-Hintergrund zu beantworten, und helfen einige Probleme bezüglich Infrarot-Divergenzen in der inflationären Kosmologie zu klären.

Acknowledgement

*»Was der Gott mich gelehrt, was mir durch's Leben geholfen,
Häng' ich, dankbar und fromm, hier in dem Heiligthum auf.«*

FRIEDRICH SCHILLER

Once Marcus Tullius Cicero said: *»There is no debt, which is more urgent to settle, than to express one's gratitude«*. On this note — thinking of all the help and patience I encountered during my doctorate — I dedicate this and the subsequent paragraphs to clearing this profound burden. As I delightfully acknowledge any kind of support — as tiny as it might have been — I have difficulties in beginning the list of those whose help and encouragement brought this work to a successful end.

I suppose my first and warmest thanks go to my parents Ruth and Rainer Kühnel. Their support, not only throughout my doctoral time, but also during my entire period of physics studies in Bielefeld and Paris, on the one hand financially and on the other hand — and vastly more important — humanly and morally, only made it possible for me to proceed that far.

Concerning the work itself, I would like to thank my advisor Prof. Dr. Dominik J. Schwarz, who always had an open ear for my problems and concerns, and who gave me the opportunity to focus on my own ideas and to do individual research with such an amount of freedom that I can only call extraordinary. Not only on the professional side, but also on the personal side, it has always been a pleasure to be a member of his research group.

I am also grateful to my second corrector Prof. Dr. Dietrich Bödeker for his appraisal of my thesis. Like my advisor, he took much time to kindly answer all of my questions and was always available for discussions, also off-topic ones...

Many many thanks go to my former and present office mates and colleagues Yannis Burnier, Benjamin Jurke, 'Fregattenkapitän' Jan Möller, and Sebastian Schmitz for sometimes very deep and controversial, but always interesting discussions, and of course for all the fun we shared. I really keep the best memories on our legendary 'Evenings of Pleasure', and I can say that they became not only colleagues but also friends.

The work in the physics department would not have been half that comfortable without a well-working and a well-organised administration. Therefore, I would like to acknowledge the support from the secretaries Mrs. Gudrun Eickmeyer and Mrs. Susi von Reder, whose kind and helpful way together with a splendiferous organisation increased the working atmosphere enormously.

I would also like to thank those, who helped to improve my thesis, either with proof-reading, constructive criticism or fruitful discussions. Among those are: Dr. Yannis Burnier, Dr. Pier-Stefano Corasaniti, Dr. Andrei Fedorenko, Dipl.-Phys. André Fischer, M.A. Prasad Hegde, Dipl.-Phys. Dipl.-Math. Benjamin Jurke, M.A. Anna Kostouki, Dipl.-Phys. Daniel Kruppke, Dr. Nán Lǐ, Dr. Jérôme Martin, Dr. Alexey Mints, Dipl.-Phys. Jan Möller, Dr. Aravind Natarajan, B.A. Catharine Oertel, Prof. Dr. Richard Woodard, Dr. Enrico Pajer, Prof. Dr. Christophe Ringeval, Dr. Aleksandar Rakić, Dr. Erandy Ramírez and Prof. Dr. Alexey Starobinsky.

Life at any stage does not flower out to its full dimension without one's friends. In this paragraph I would like to deeply acknowledge the support of all those who made my doctorate so nice as it actually turned out. As I already expressed my gratitude to many friends of mine, I would now additionally like to thank Lena Bloemertz, Thorsten and Tobias Bogner, Marciana-Nona Duma, Lars Heidieker, Julia Hertle and Thea Krauß.

Last but not least, my especial thanks shall be due to my very dear grandparents Erich Häusler, Anneliese Häusler and Anni Kühnel, my brother Christian, my grandaunt Margot Brinkmann, and Hans Wallrafen.

Contents

1. Cosmological Inflation *11*
   1.1 Friedmann Space-Time *11*
   1.2 Shortcomings of the Standard Big Bang Theory *14*
      1.2.1 Flatness Problem *14*
      1.2.2 Horizon Problem *14*
   1.3 Standard Inflationary Universe *15*
      1.3.1 Slow-Roll *16*
      1.3.2 Inflation and the Shortcomings of the Standard Big Bang Theory *17*
   1.4 Inflation and Cosmological Perturbations *18*

2. Cosmological Perturbations & Primordial Inhomogeneities *19*
   2.1 Perturbations *20*
   2.2 Gauges *21*
   2.3 Gauge-Invariant Variables *22*
   2.4 Dynamical Equations *23*
   2.5 Hydro-Dynamical Perturbations *24*
      2.5.1 Long-Wavelength Solutions *25*
      2.5.2 Short-Wavelength Solutions *26*
   2.6 Baryon-Radiation Plasma and Cold Dark Matter *26*
      2.6.1 Long-Wavelength Solutions *27*
      2.6.2 Short-Wavelength Solutions *28*
   2.7 Origin of Primordial Inhomogeneities *29*
      2.7.1 Perturbations *30*
      2.7.2 Canonical Scalar Field *31*
      2.7.3 Inflationary Gravitational Waves *32*

3. Cosmic Microwave Background Anisotropies *35*
   3.1 Boltzmann Equation and Temperature Fluctuations *36*
   3.2 Sachs-Wolfe Effect *38*
   3.3 Temperature Correlation *41*
   3.4 Other Effects *43*
      3.4.1 Small-Scale Anisotropies *44*

	3.4.2 Reionisation	46
	3.4.3 Gravitational Waves	47
	3.4.4 Polarisation	48

4 Stochastic Cosmological Inflation — 51
    4.1 Effective Equation of Motion — 52
    4.2 Other Results — 57

5 Replica Field Theory — 61
    5.1 Random Field Ising Model — 62
    5.2 Elastic Systems — 64
    5.3 Replica Trick — 65
    5.4 Noise Distributions — 67
    5.5 Variational Method — 68
        5.5.1 Feynman-Jensen Inequality — 68
        5.5.2 Long-Range Correlation — 71
        5.5.3 Replica Symmetric Propagator — 72

6 Stochastic Inflation & Replica Field Theory — 73
    6.1 Free Power Spectrum — 74
    6.2 Effective Power Spectrum — 76
        6.2.1 Long-Range Correlation — 76
        6.2.2 Stochastic Inflation — 77
    6.3 Filter Functions — 80
        6.3.1 Modified Gaussian Fluctuations — 88
    6.4 Self-Interactions — 91
    6.5 Possible CMB effects — 92

Summary & Outlook — 97

Appendix: Replica Symmetry Breaking — 99

Bibliography — 109

Chapter 1
Cosmological Inflation

»ἁρμονίη ἀφανὴς φανερῆς κρείττων.«
[»*The hidden harmony is better than the open.*«]

HERACLITUS OF EPHESUS

A consistent and fundamental understanding of the observed isotropy of our Universe is one the main goals of modern cosmology. Based on old ideas of Starobinsky [Sta80] and Guth [Gut80], the Universe underwent a phase of quasi-exponential expansion, **inflation**, before the time when it was radiation dominated. As powerful as it is, this set-up can — at least phenomenologically — be realised in an extremely simple manner in terms of a single scalar field in a sufficiently flat potential, causing the desired expansion. The quantum fluctuations of this field, called **inflaton field**, as they are stretched to cosmological scales, are responsible for the observed fluctuations in the cosmic microwave background radiation and provide the seeds for the formation of structure in general.

1.1 Friedmann Space-Time

Before we discuss the usefulness of the idea of cosmological inflation, we first present some basics of standard Big Bang cosmology. These insights allow us to understand several of its conceptual insufficiencies. We will restrict ourselves to a rather brief discussion as most of these issues are well-known in cosmology and shall serve more as a remainder of the most important concepts and formulae, as well as to fix notation.

We start with the basic four-dimensional space-time accounting for the observed isotropy in the matter distribution on a few hundred megaparsecs: The **Friedmann Universe**. This space-time can be defined by the line element

$$\mathrm{d}s^2 = \mathrm{g}_{\mu\nu}\mathrm{d}x^\mu\mathrm{d}x^\nu = \mathrm{d}t^2 - a(t)^2\left[\frac{\mathrm{d}r^2}{1-\kappa r^2} + r^2\left(\mathrm{d}\theta^2 + \sin^2(\theta)\mathrm{d}\phi^2\right)\right], \quad (1.1)$$

with the **metric** $\mathrm{g}_{\mu\nu}$, the **scale factor** $a(t)$, and the constant κ which can take the values $1, -1, 0$, depending on whether the Universe is closed, open or flat. Observations suggest [KDN+09] that the latter is realised, and hence we will mainly consider this case in the following. In (1.1) and throughout this work we employed the **Einstein summation convention**, i.e. any expression with two common indices implicitly means a sum over all of their values: Greek indices run from 0 to 3 and Latin over 1, 2 and 3.

The evolution of the scale factor can be calculated by the **Einstein field equations**

$$\mathrm{G}_{\mu\nu} = \mathrm{T}_{\mu\nu}, \quad (1.2)$$

with the **Einstein tensor** $\mathrm{G}_{\mu\nu}$, the **energy-momentum tensor** $\mathrm{T}_{\mu\nu}$, and for convenience we set $8\pi G \stackrel{!}{=} 1$, $\hbar \stackrel{!}{=} 1$, and $c \stackrel{!}{=} 1$. With the assumptions of homogeneity and isotropy, the energy-momentum tensor can be written in the diagonal form $(\mathrm{T}^\mu{}_\nu) = \mathrm{diag}(\varepsilon, -p, -p, -p)$, which yields from the 0-0-component of (1.2) the **Friedmann equation**

$$H^2 \equiv \left(\frac{\dot{a}}{a}\right)^2 = \frac{\varepsilon}{3} - \frac{\kappa}{a^2}. \quad (1.3)$$

Equation (1.3) also defines the **Hubble parameter** H, the basic quantity measuring the expansion rate, where as usual, a dot means a differentiation with respect to time. One then finds from (1.2)

$$\dot{H} = -\frac{1}{2}(\varepsilon + p) + \frac{\kappa}{a^2}. \quad (1.4)$$

The Friedmann equation (1.3) provides a relation of the curvature of the Universe to the energy density and the expansion rate via

$$\Omega - 1 = \frac{\kappa}{a^2 H^2}, \quad (1.5)$$

where $\Omega := \varepsilon/\varepsilon_{\mathrm{crit.}}$ with $\varepsilon_{\mathrm{crit.}} := 3H^2$. Now, if $\Omega > 1$ the Universe is positively curved, $\Omega < 1$ corresponds to a negative curvature and finally $\Omega = 1$ implies a flat space-time.

This last case seems to be realised to good precision, as combined measurements give $\Omega_0 := \Omega(t_0) = 1.00^{+0.07}_{-0.03}$ [KDN+09]. The present understanding is that Ω_0 is a sum of mainly three parts: $\Omega_0 = \Omega_b + \Omega_{\mathrm{DM}} + \Omega_\Lambda$ (baryons, dark matter, dark energy), where these fractions have been determined

1.1 Friedmann Space-Time

to $\Omega_b = 0.044 \pm 0.003$, $\Omega_{DM} = 0.214 \pm 0.027$ and $\Omega_\Lambda = 0.742 \pm 0.030$; the radiation fraction is negligible today [KDN+09].

With the so-called **conformal time** η, defined via $d\eta \equiv a(t)^{-1} dt$, and for $\kappa = 0$, the metric $g_{\mu\nu}$ underlying the line element (1.1) can be written as

$$g_{\mu\nu} = a(\eta)^2 \eta_{\mu\nu}, \tag{1.6}$$

with the Minkowski metric $\eta_{\mu\nu}$. This implies for a **power-law scale factor** $a(t) \propto t^n$ the result $a(\eta) \propto \eta^{n/(1-n)}$. An **exponential scale factor** $a(t) \propto e^{Ht}$ yields $a(\eta) \propto \eta^{-1}$, which can be obtained from the former case in the limit $n \to \infty$. In general, for an **equation of state** $p/\varepsilon = \omega \in \mathbb{R}$, one finds $a(t) \propto t^{2/(3+3\omega)}$ and $\varepsilon \propto a^{-3-3\omega}$. If the Universe is **matter dominated** one has $\omega = 0$, implying $a(t) \propto t^{2/3}$ and $\varepsilon \propto a^{-3}$. The case of **radiation domination** is described by $\omega = 1/3$, which yields $a(t) \propto t^{1/2}$, $\varepsilon \propto a^{-4}$, and so $a \propto T^{-1}$ if $\varepsilon \propto T^4$, where T is the temperature of the Universe.

It is convenient to define analogous to H the quantity $\mathcal{H} := a'/a$, where a prime denotes differentiation with respect to η. Then equation (1.4) becomes

$$\mathcal{H}' = -\frac{1}{6}(\varepsilon + 3p). \tag{1.7}$$

Other useful relations are

$$H = \frac{\mathcal{H}}{a}, \quad \dot{H} = \frac{\mathcal{H}' - \mathcal{H}^2}{a^2}, \quad \mathcal{H}^2 = \frac{\varepsilon a^2}{3} - \kappa. \tag{1.8}$$

A concept we will employ is that of a **particle horizon** R_H. It is defined as the maximal distance a photon can travel since the Big Bang (at time t_i) until time t:

$$R_H(t) = a(t) \int_{t_i}^{t} \frac{dt'}{a(t')} = a(\eta) \int_{\eta_i}^{\eta} d\eta'. \tag{1.9}$$

One finds

$$R_H = \begin{cases} \frac{n}{n-1}\frac{1}{H}\left[(t/t_i)^{n-1} - 1\right] & \text{(power-law)}, \\ \frac{1}{H}\left[e^{H(t-t_i)} - 1\right] & \text{(exponential)}. \end{cases} \tag{1.10}$$

We see that, while H is constant in the exponential case, R_H diverges if $t_i \to -\infty$, meaning that all points were in causal contact. However, one should keep in mind that the de Sitter space-time is geodesically incomplete and is only used as an approximation.

1.2 Shortcomings of the Standard Big Bang Theory

Having briefly reviewed some basic concepts of the standard Big Bang cosmology in the previous section, we will now discuss two of its insufficiencies, namely the so-called flatness problem and the horizon problem. Although termed 'problems', these issues do not reflect any logical inconsistency, but require a rather fine-tuned 'unnatural' arrangement of the initial conditions to describe the measured cosmological data.

1.2.1 Flatness Problem

We have already mentioned the observed flatness of the present Universe. If today $\Omega_0 \simeq 1$, one may ask what are the implications for the initial conditions of the Universe. To this end we notice that in the case of radiation domination

$$\Omega - 1 \propto a^2 \,, \tag{1.11}$$

which implies with the temperature dependence $a \propto T^{-1}$ for the ratio

$$\frac{|\Omega - 1|_{t=t_i}}{|\Omega - 1|_{t=t_0}} \simeq \left(\frac{a(t_i)}{a(t_0)}\right)^2 \simeq \left(\frac{T_0}{T_i}\right)^2 . \tag{1.12}$$

As the present-day temperature of the cosmic microwave background photons is roughly $T_0 \approx 10^{-13}\,\text{GeV}$, we find that the choice of $T_i = T_{\text{Planck}} = 10^{19}\,\text{GeV}$ yields for the right-hand side of (1.12) the value 10^{-64}. Thus, to get the observed value of Ω_0 one needs an amazingly strong fine-tuning.

1.2.2 Horizon Problem

As mentioned, the Universe today is extremely homogeneous on scales of a few hundred megaparsecs. This domain of homogeneity is at least as large as the present horizon, which represents the maximal length at which causal processes can take place. Its linear extension has initially been smaller by the ratio of the corresponding scale factors a_i/a_0, implying that the size of the homogeneous and isotropic patch from which our Universe originated was at least as large as (linear extension)

$$\ell_i \sim t_0 \frac{a_i}{a_0} \,, \tag{1.13}$$

assuming that inhomogeneity cannot be dissolved by expansion.

If one compares this to the size of a causal region of linear extent ℓ_c (at initial time t_i), one obtains

$$\frac{\ell_i}{\ell_c} \sim \frac{t_0}{t_i} \frac{a_i}{a_0} . \tag{1.14}$$

Hence, assuming radiation domination and $t_i = t_\text{Planck}$, we find the ratio to be 10^{28}, meaning that the Universe was very smooth over $(10^{28})^3 = 10^{84}$ initially causally disconnected regions. Therefore, there is no causal physical process in a decelerating Friedmann Universe that can be responsible for this extremely fine-tuned matter distribution.

1.3 Standard Inflationary Universe

The previous discussion of the shortcomings of the Big Bang theory suggests that the Universe underwent a period in which the horizon scale H^{-1} grew slower than a physical scale λ_ph. As these grow like $a(t)$, the previous consideration yields

$$0 < \frac{\mathrm{d}}{\mathrm{d}t}\left(\frac{\lambda_\text{ph}}{H^{-1}}\right) \propto \ddot{a}, \tag{1.15}$$

as a criterion for **cosmological inflation**, i.e. *accelerated expansion*. It can be deduced from equation (1.7) that

$$p < -\frac{\varepsilon}{3} \tag{1.16}$$

is required to have $\ddot{a} > 0$ and therefore neither radiation- nor matter-domination fulfills this requirement.

One remarkably simple possibility to realise this is by means of a single scalar field φ, called **inflaton**. Let us for simplicity assume its action to be the **canonical** one,

$$\mathcal{S}[\varphi] = \int \mathrm{d}^4 x\, \sqrt{-g}\, \mathcal{L}[\varphi] = \int \mathrm{d}^4 x\, \sqrt{-g}\, \left[\frac{1}{2} g^{\mu\nu} \varphi_{,\mu} \varphi_{,\nu} - V(\varphi)\right], \tag{1.17}$$

for which the corresponding energy-momentum tensor $T_{\mu\nu}$ reads

$$T_{\mu\nu} = \frac{2}{\sqrt{-g}} \frac{\delta \mathcal{S}[\varphi]}{\delta g^{\mu\nu}} = \varphi_{,\mu} \varphi_{,\nu} - g_{\mu\nu} \mathcal{L}[\varphi]. \tag{1.18}$$

This implies

$$\varepsilon = T^0{}_0 = \frac{\dot{\varphi}^2}{2} + V(\varphi) + \frac{(\nabla \varphi)^2}{2 a^2}, \tag{1.19a}$$

$$p = -\frac{1}{3} T^i{}_i = \frac{\dot{\varphi}^2}{2} - V(\varphi) - \frac{(\nabla \varphi)^2}{6 a^2}. \tag{1.19b}$$

Hence, we find $p = -\varepsilon/3$ if the gradient term dominates, if the potential dominates one has $p = -\varepsilon$, and $p = \varepsilon$ if the kinetic term prevails.

1.3.1 Slow-Roll

Let us split the scalar field into a homogeneous part φ_0 and fluctuations $\delta\varphi$ around φ_0,

$$\varphi(t, \boldsymbol{x}) = \varphi_0(t) + \delta\varphi(t, \boldsymbol{x}) \, . \tag{1.20}$$

Since fluctuations will intensively be discussed the next chapters, we will focus here on the homogeneous part φ_0. It obeys the equation of motion

$$\ddot{\varphi}_0 + 3H\dot{\varphi}_0 + \frac{\mathrm{d}V(\varphi_0)}{\mathrm{d}\varphi_0} = 0 \, . \tag{1.21}$$

The so-called **slow-roll** approximation consists of taking $\dot{\varphi}_0^2 \ll V(\varphi_0)$ and $|\ddot{\varphi}_0| \ll |H\dot{\varphi}_0|$, and assuming that the potential is sufficiently flat. It reduces equation (1.21) to

$$3H\dot{\varphi}_0 \simeq -\frac{\mathrm{d}V(\varphi_0)}{\mathrm{d}\varphi_0} \, . \tag{1.22}$$

Then, the assumptions that the inflaton field dominates the energy density yields

$$H^2(\varphi_0) \simeq \frac{1}{3}V(\varphi_0) \, , \tag{1.23}$$

and hence that the slow-roll condition can be expressed in terms of constraints on the **first slow-roll parameter**

$$\epsilon_{\mathrm{sr}} := -\frac{\mathrm{d}\ln(H)}{\mathrm{d}\ln(a)} = -\frac{\dot{H}}{H^2} = \frac{\dot{\varphi}_0^2}{2H^2} \simeq \frac{1}{2}\left(\frac{V'}{V}\right)^2 \ll 1 \, , \tag{1.24a}$$

and the **second slow-roll parameter**

$$\eta_{\mathrm{sr}} := -\frac{\mathrm{d}\ln(H')}{\mathrm{d}\ln(a)} \simeq \frac{V''}{V} \simeq \frac{V''}{3H^2} \ll 1 \, . \tag{1.24b}$$

Above, a prime denotes a derivative with respect to φ_0. From the definition of ϵ_{sr} it follows that

$$\frac{\ddot{a}}{a} = H^2 + \dot{H} = (1 - \epsilon_{\mathrm{sr}})H^2 \, , \tag{1.25}$$

and hence inflation can only occur for $\epsilon_{\mathrm{sr}} < 1$. Furthermore, we have $\dot{\epsilon}_{\mathrm{sr}}, \dot{\eta}_{\mathrm{sr}} = \mathcal{O}(\epsilon_{\mathrm{sr}}^2, \eta_{\mathrm{sr}}^2)$, implying that working to first order in the slow-roll parameters tantamounts to take them as constants in time.

Usually, the amount of inflation is quantified by the logarithm of the ratio of the scale factor at the end of inflation at time t_f, to its value at some time t. It is called **number of e-folds**

$$N(t) := \ln\left(\frac{a(t_f)}{a(t)}\right) . \tag{1.26}$$

1.3 Standard Inflationary Universe

In the case of single-field slow-roll inflation one has

$$N(t) = \int_t^{t_f} dt'\, H(t') \simeq -\int_{\varphi(t)}^{\varphi(t_f)} d\varphi\, \frac{V(\varphi)}{V'(\varphi)}\,. \tag{1.27}$$

Let us once again return to the equation of state for the scalar-field model under consideration, i.e. to $p = -\varepsilon$. From the equations (1.3) and (1.4) we find that in this case both ε and H are constant. This implies for the scale factor

$$a(t) = a_i\, e^{H(t-t_i)}\,, \tag{1.28}$$

and thus the total number of e-folds is just given by

$$N_{\scriptscriptstyle\rm T} := N(t_i) = H(t_f - t_i)\,. \tag{1.29}$$

1.3.2 Inflation and the Shortcomings of the Standard Big Bang Theory

We will first discuss how inflation can provide a *solution to the horizon problem*. As mentioned previously, the (de Sitter) horizon H is constant. Thus, for sufficiently many e-folds all physical scales that left the horizon could have been subhorizon, because of their exponential suppression in the past.

Let us therefore calculate the number of e-folds which are necessary to solve the horizon problem. For simplicity we assume a two-stage history of the Universe with an instantaneous transition from an exponential to a radiation-dominated period.

The minimum requirement is that a scale as large as the present horizon H_0^{-1} has been inside the horizon at the beginning of inflation, i.e. inside H_i^{-1}. Hence,

$$H_i^{-1} \gtrsim H_0^{-1}\, \frac{a(t_i)}{a(t_0)} = H_0^{-1}\, \frac{a(t_f)}{a(t_0)}\, \frac{a(t_i)}{a(t_f)} \simeq H_0^{-1}\, \frac{T_0}{T_f}\, e^{-N_{\scriptscriptstyle\rm T}} \tag{1.30}$$

and thus for the choice $H_0 = 10^{-42}$ GeV and $T_0 = 10^{-13}$ GeV the lower bound

$$N_{\scriptscriptstyle\rm T} \gtrsim \ln\!\left(\frac{T_0}{H_0}\right) - \ln\!\left(\frac{T_f}{H_i}\right) \simeq 67 - \ln\!\left(\frac{T_f}{H_i}\right), \tag{1.31}$$

where T_f is the temperature at the end of inflation, just before radiation-domination.

The size of the last term in (1.31) is very model dependent and determined by the value of the temperature T_f at so-called reheating and the initial Hubble constant H_i. In the case of **chaotic inflation**, where all relevant energy scales are Planckian, and for $T_f = T_{\scriptscriptstyle\rm GUT} \approx 10^{16}$ GeV we find

$\ln(H_i/T_f) \approx 7$, whereas for $T_f = T_{\text{EW}} \approx 10^2\,\text{GeV}$ the corresponding value is more than five times larger. For a discussion on the issue of the number of e-folds, we refer interested reader to the article [LL03].

Inflation also elegantly *solves the flatness problem*. Because H is constant during inflation, we find

$$\Omega - 1 \propto a^{-2}, \qquad (1.32)$$

which implies

$$\frac{|\Omega - 1|_{t=t_f}}{|\Omega - 1|_{t=t_i}} \approx \left(\frac{a(t_i)}{a(t_f)}\right)^2 = e^{-2N_T}. \qquad (1.33)$$

This means that if inflation lasts long enough, Ω can be made as close to one as desired. Of course, inflation-induced flatness is to be understood cum grano salis: A space-time with an open or closed geometry will stay open or closed, respectively. It is only that the observable patch becomes much smaller than the curvature radius of the Universe.

1.4 Inflation and Cosmological Perturbations

Although being intensively discussed in the next chapter, we will now briefly comment on the link of inflation to the generation of cosmological perturbations, for completeness.

Inflation may provide a simple and almost perfect way for homogenisation and isotropisation our observable patch of the Universe. Now, 'almost' is a key to the issue of structure formation. In fact, the most important aspect of inflation is the explanation of the origin of large-scale structure, which, as discussed, could not be generated by causal processes, since, e.g. the angular distance of the horizon at radiation/matter equality is about one degree.

What basically happens is that quantum fluctuations of the scalar field induce fluctuations in the gravitational potential. This leads in turn to a concentration of matter in the gravitational troughs, which attracts more and more matter. The quasi-exponential inflation is then stretching these seeds to cosmological scales.

Chapter 2

Cosmological Perturbations
&
Primordial Inhomogeneities

»Anbeginn von Raum und Zeiten –
urgetrieben Weltenkeim
lässt den Kosmos fortan schreiten;
Schau, ein Stern – wir sind daheim.«

F. K.

At the time of recombination, the Universe was very isotropic and homogeneous. Also today, when averaged over a few hundred megaparsecs, inhomogeneities in the density distributions are small. However, with galaxies and clusters, superclusters or filaments of such, the Universe has a well developed non-linear structure. An explanation for this is found in **gravitational instabilities** originating from the attractive nature of gravity. Over-dense regions attract matter, which increases the density further and, in turn, leads to an amplification of the original inhomogeneity in the distribution of matter.

In this chapter, which follows, and in part summarises, [Muk05], we will first discuss the issue of metric perturbations and gravitational instabilities. Afterwards we apply this to various types of perturbations and study the origin of primordial inhomogeneities, including their spectrum. We conclude with a discussion of quantum cosmological perturbations.

2.1 Perturbations

Let us study small perturbations δg around a flat Friedmann Universe using the metric

$$ds^2 = g_{\alpha\beta}\, dx^\alpha\, dx^\beta \equiv \left(g^{(0)}{}_{\alpha\beta} + \delta g_{\alpha\beta}\right) dx^\alpha\, dx^\beta\,, \tag{2.1}$$

where, $g^{(0)}{}_{\alpha\beta}(\eta) = a(\eta)^2\, \eta_{\alpha\beta}$ with $|\delta g_{\alpha\beta}| \ll |g^{(0)}{}_{\alpha\beta}|$, for all α, β. In (2.1) and throughout this work, a superscript '(0)' shall indicate *background* quantities. Since $(\delta g_{\alpha\beta})$ is real and symmetric, it has 10 real, independent components. It is convenient to categorise them into *scalar*, *vector* and *tensor* perturbations. To see how this precisely works, we write — without loss of generality — the metric fluctuations in the form

$$\delta g_{00} \equiv 2\, a^2\, \phi\,, \tag{2.2a}$$

$$\delta g_{0i} \equiv a^2\left(B_{,i} + S_i\right), \tag{2.2b}$$

$$\delta g_{ij} \equiv a^2\left(2\,\psi\,\delta_{ij} + 2\,E_{,ij} + 2\,F_{i,j} + 2\,F_{j,i} + 2\,h_{ij}\right), \tag{2.2c}$$

with the constraints that the 3-vectors $\vec{F}$ and $\vec{S}$ are *divergence-free*

$$F^i{}_{,i} = 0\,, \qquad S^i{}_{,i} = 0\,, \tag{2.3}$$

and that the 3-tensor h is *traceless* and *transverse*,

$$h^i{}_i = 0\,, \qquad h^{ij}{}_{,i} = 0\,. \tag{2.4}$$

Now, **scalar perturbations** are described by the four functions ϕ, ψ, B and E. They exhibit gravitational instabilities and are thus considerably responsible for the formation of large-scale structure.

Vector perturbations are characterised by the two 3-vectors $\vec{F}$ and $\vec{S}$, having together four independent components due to the constraints (2.3). Because they decay quickly — since they are related to rotational motion, which is damped in an expanding Universe — they are of minor interest in inflationary cosmology.

The **tensor perturbations** describe **gravitational waves**, which are encoded in the transverse and traceless 3-tensor h, that has two independent components, regarding the constraints (2.4). To linear order, they do not induce any perturbations in the baryon-radiation fluid.

A trivial, but important fact is that these kinds of perturbations *do not mix* at linear order and hence can be studied separately. In the following paragraph we will *focus on scalar perturbations*.

2.2 Gauges

Having now classified the different types of perturbations, we will next classify different gauges. With a **gauge transformation** we shall understand a general coordinate transformation

$$x \to \tilde{x}(x),\qquad(2.5)$$

where we restrict ourselves to *infinitesimal* ones, i.e.

$$\tilde{x} = x + \zeta,\qquad(2.6)$$

with ζ chosen such that it leads to infinitesimal changes in observables only.

Let us briefly discuss two important choices of (classes of) coordinate systems. The first is the so-called **longitudinal** (or **conformal-Newton**) **gauge**, which is *fixed* by

$$B \equiv E \equiv 0.\qquad(2.7)$$

With this choice, the line element takes the form

$$\mathrm{d}s^2 = a^2 \left[(1+2\phi)\,\mathrm{d}\eta^2 - (1-2\psi)\,\mathrm{d}\boldsymbol{x}\cdot\mathrm{d}\boldsymbol{x}\right].\qquad(2.8)$$

As shown below in this chapter, for a diagonal spatial part of the energy-momentum tensor, i.e. with $T_{ij} \propto \delta_{ij}$, one has $\phi = \psi$. Furthermore, the variable ϕ is a *generalisation of the Newton potential*, justifying thus the name conformal-*Newton* gauge.

The **synchronous gauge** is defined by the choice

$$\phi \equiv B \equiv 0,\qquad(2.9\mathrm{a})$$

or equivalently,

$$\delta g_{0\alpha} \equiv 0.\qquad(2.9\mathrm{b})$$

We should stress that this gauge does not specify a certain observer but rather a *class* of coordinate systems, due to the residual gauge freedom

$$\eta \to \tilde{\eta}(\eta, \boldsymbol{x}) := \eta + \frac{c_1(\boldsymbol{x})}{a(\eta)},\qquad(2.10\mathrm{a})$$

$$\boldsymbol{x} \to \tilde{\boldsymbol{x}}(\eta, \boldsymbol{x}) := \boldsymbol{x} + \left(\nabla_{\boldsymbol{x}} c_1(\boldsymbol{x})\right)\int \frac{\mathrm{d}\eta}{a(\eta)} + \nabla_{\boldsymbol{x}} c_2(\boldsymbol{x}),\qquad(2.10\mathrm{b})$$

which leaves (2.9a,b) invariant. c_1 and c_2 are arbitrary functions of $\boldsymbol{x}$ only.

Hence, the physical interpretation in this gauge might be difficult due to the appearance of unphysical gauge modes. It is therefore common practice to supplement the synchronous gauge with another condition to fix the residual freedom, e.g. the comoving synchronous gauge.

Appart from these two, there is a full variety of other gauges. All of those are defined by certain choices of (classes of) coordinate systems, which themselves are connected by diffeomorphisms. According to this, one has a defined link between the various metric perturbations.

It seems necessary to remark further on the word *gauge*: Usually in field theory, a gauge refers to a transformation of the associated vector potential within the inherent redundancy of such a description. This has nothing to do with the coordinate system, whereas a gauge in general relativity specifies exactly that: A particular observer. Hence, every measurement *selects* a certain gauge.

2.3 Gauge-Invariant Variables

Let us now derive connections between different gauges. Therefore we take the coordinate transformation (2.6) and write

$$\vec{\zeta} = \vec{\zeta}_\perp + \nabla_{\bm{x}} \xi \,, \tag{2.11}$$

where $\nabla_{\bm{x}} \cdot \vec{\zeta}_\perp = 0$ and ξ is a scalar function. Then we find

$$\delta \tilde{g}_{00} = \delta g_{00} - 2a \left(a \, \zeta^0 \right)' \,, \tag{2.12a}$$

$$\delta \tilde{g}_{0i} = \delta g_{0i} + a^2 \left[(\zeta_\perp)'_i + \left(\xi' - \zeta^0 \right)_{,i} \right] \,, \tag{2.12b}$$

$$\delta \tilde{g}_{ij} = \delta g_{ij} + a^2 \left[2 \frac{a'}{a} \zeta^0 \delta_{ij} + 2 \, \xi_{,ij} + \left((\zeta_\perp)_{i,j} + (\zeta_\perp)_{j,i} \right) \right] \,, \tag{2.12c}$$

where a comma denotes a partial derivative with respect to the corresponding coordinate. For scalar perturbations, the line element takes the form

$$\mathrm{d}s^2 = a^2 \left[(1 + 2\,\phi) \, \mathrm{d}\eta^2 + 2\,\mathrm{B}_{,i} \, \mathrm{d}x^i \, \mathrm{d}\eta - \left((1 - 2\,\psi) \delta_{ij} - 2\,\mathrm{E}_{,ij} \right) \mathrm{d}x^i \, \mathrm{d}x^j \right] \,, \tag{2.13}$$

which implies for the transformation of the scalar metric functions:

$$\phi \to \tilde{\phi} = \phi - \frac{1}{a} \left(a \, \zeta^0 \right)' \,, \tag{2.14a}$$

$$\psi \to \tilde{\psi} = \psi + \frac{a'}{a} \zeta^0 \,, \tag{2.14b}$$

$$\mathrm{B} \to \tilde{\mathrm{B}} = \mathrm{B} + \xi' - \zeta^0 \,, \tag{2.14c}$$

$$\mathrm{E} \to \tilde{\mathrm{E}} = \mathrm{E} + \xi \,. \tag{2.14d}$$

2.4 Dynamical Equations

This allows one to construct **gauge-independent** quantities, i.e. those that are *invariant under coordinate changes*. From (2.14a-d) one sees that (for scalar perturbations) only ζ^0 and ξ contribute to the transformation and hence one can make any two of ϕ, ψ, B, or E vanish. Two simple gauge-invariant linear combinations of these quantities are

$$\Phi := \phi - \frac{1}{a}\left[a\left(B - E'\right)\right]', \tag{2.15a}$$

$$\Psi := \psi + \frac{a'}{a}\left(B - E'\right). \tag{2.15b}$$

A nice feature about these variables is that, by construction, they do not change under coordinate transformations and hence, if one of them is zero in a certain frame, it is zero for any observer. In particular, if both Φ and Ψ are zero, there is no physical scalar metric perturbation. Furthermore, if one has a solution to the Einstein field equations derived in one gauge, one can easily express it in any gauge using gauge-invariant variables, without solving these equations again.

2.4 Dynamical Equations

To derive dynamical equations for the metric perturbations, we employ the Einstein field equations,

$$G^{(0)\alpha}{}_{\beta} + \delta G^{\alpha}{}_{\beta} \equiv G^{\alpha}{}_{\beta} = T^{\alpha}{}_{\beta} \equiv T^{(0)\alpha}{}_{\beta} + \delta T^{\alpha}{}_{\beta}, \tag{2.16}$$

which imply for the corresponding linear part in the metric perturbations

$$\delta G^{\alpha}{}_{\beta} = \delta T^{\alpha}{}_{\beta}. \tag{2.17}$$

It is important to note that *neither δG nor δT are gauge invariant*, but are easily modified to yield gauge-invariant objects. For the linear fluctuations of the energy-momentum tensor one finds that

$$\delta \bar{T}^0{}_0 := \delta T^0{}_0 - \left(T^{(0)0}{}_0\right)'\left(B - E'\right), \tag{2.18a}$$

$$\delta \bar{T}^0{}_i := \delta T^0{}_i - \left(T^{(0)0}{}_0 - \frac{1}{3}T^{(0)k}{}_k\right)\left(B - E'\right)_{,i}, \tag{2.18b}$$

$$\delta \bar{T}^i{}_j := \delta T^i{}_j - \left(T^{(0)i}{}_j\right)'\left(B - E'\right) \tag{2.18c}$$

provides a gauge-independent version, and for the fluctuations of the Einstein tensor one has similiar expressions but with T replaced by G. Scalar perturbations imply after some algebra (c.f. [Muk05])

$$\Delta\Psi - 3\mathcal{H}\left(\Psi' + \mathcal{H}\Phi\right) = \frac{1}{2}a^2\,\delta\bar{T}^0{}_0\,, \tag{2.19a}$$

$$\left(\Psi' + \mathcal{H}\Phi\right)_{,i} = \frac{1}{2}a^2\,\delta\bar{T}^0{}_i\,, \tag{2.19b}$$

$$\left[\Psi'' + \mathcal{H}\left(2\Psi + \Phi\right)' + \left(2\mathcal{H}' + \mathcal{H}^2\right)\Phi \right. \\ \left. + \frac{1}{2}\Delta(\Phi - \Psi)\right]\delta^i{}_j - \frac{1}{2}(\Phi - \Psi)^{,i}{}_{,j} = -\frac{1}{2}a^2\,\delta\bar{T}^i{}_j\,, \tag{2.19c}$$

from which we see that the spatial part of the energy-momentum tensor is diagonal if and only if $\Phi \equiv \Psi$.

2.5 Hydro-Dynamical Perturbations

Having now equations for scalar perturbations for an arbitrary energy-momentum tensor, we next discuss their solutions for special choices of $T^\mu{}_\nu$. The first will be that of a **perfect fluid**, which is described by

$$T^\mu{}_\nu = (\varepsilon + p)\,u^\mu u_\nu - p\,\delta^\mu{}_\nu \tag{2.20}$$

with $u^\mu u_\mu = 1$. It implies for the corresponding linear fluctuations,

$$\delta\bar{T}^0{}_0 = \delta\bar{\varepsilon} := \delta\varepsilon - \varepsilon_0'\left(B - E'\right), \tag{2.21a}$$

$$\delta\bar{T}^0{}_i = \frac{1}{a}(\epsilon_0 + p_0)\,\delta\bar{u}_i \equiv \frac{1}{a}(\epsilon_0 + p_0)\left[\delta u_i - a\left(B - E'\right)_{,i}\right], \tag{2.21b}$$

$$\delta\bar{T}^i{}_j = -\delta\bar{p}\,\delta^i{}_j \equiv \left[\delta p - p_0'\left(B - E'\right)\right]\delta^i{}_j\,. \tag{2.21c}$$

Thus, as stated before, one finds from (2.19c) and the diagonal structure of (2.20) that the potentials Φ and Ψ are equal.

The equation we would like to derive is one which characterises the dynamics of Φ. To this end, we decompose the pressure fluctuation $\delta\bar{p}$ into fluctuations of the energy density ε and the entropy S according to

$$\delta\bar{p} = \left.\frac{\partial p}{\partial \varepsilon}\right|_S \delta\bar{\varepsilon} + \left.\frac{\partial p}{\partial S}\right|_\varepsilon \delta S \equiv c_s^2\,\delta\bar{\varepsilon} + \rho\,\delta S\,, \tag{2.22}$$

which defines the speed of sound c_s and the parameter ρ. One can then combine the equations (2.19a-c) and (2.21a-c) to the desired *single* equation for Φ,

$$\Phi'' + 3\left(1 + c_s^2\right)\mathcal{H}\,\Phi' - c_s^2\,\Delta\Phi + \left[2\mathcal{H}' + \left(1 + 3c_s^2\right)\mathcal{H}^2\right]\Phi = \frac{1}{2}a^2\rho\,\delta S\,. \tag{2.23}$$

2.5 Hydro-Dynamical Perturbations

In the following paragraph we will focus on **adiabatic perturbations**, where $\delta S \equiv 0$. To simplify equation (2.23) we first eliminate the "friction" term proportional to Φ' by the substitution

$$v := \frac{\Phi}{\sqrt{\varepsilon_0 + p_0}} . \tag{2.24}$$

Then, with $\omega_0 := p_0/\varepsilon_0$ and the definition

$$\vartheta := \frac{1}{a}\frac{1}{\sqrt{1+\omega_0}} = \frac{1}{a}\left[\frac{2}{3}\left(1 - \frac{\mathcal{H}'}{\mathcal{H}^2}\right)\right]^{-1/2} , \tag{2.25}$$

one arrives after some algebra at (c.f. [Muk05])

$$v'' - c_s^2 \Delta v - \frac{\vartheta''}{\vartheta} v = 0 . \tag{2.26}$$

2.5.1 Long-Wavelength Solutions

For the regime we are primarily interested in (c.f. the discussion of the next chapter), the **long-wavelength perturbations**, i.e. those with $c_s \eta k \ll 1$, we can neglect spatial derivatives in (2.26) and obtain the first solution

$$v(\eta) \propto \vartheta(\eta) \int_{\eta_0}^{\eta} \frac{d\tau}{\vartheta^2(\tau)} . \tag{2.27}$$

The second solution is proportional to ϑ and decays with time (c.f. (2.25)).

From the long-wavelength solution (2.27) it is easy to construct a *conserved* quantity. The simplest possibility is (c.f. [Muk05])

$$\vartheta^2 \left(\frac{v}{\vartheta}\right)' = \sqrt{3}\, \frac{\mathcal{H}^{-1}\Phi' + \Phi}{1+\omega} + \Phi , \tag{2.28}$$

which is constant even if ω is changing.

Let us now assume that the system consists of two components with different constant equation of state, $p_{i/f} = \omega_{i/f}\,\varepsilon_{i/f}$, for which one dominates at conformal time η_i and the other at η_f. Then the constancy of (2.28) shows (c.f. [Muk05])

$$\Phi(\eta_f) = \frac{1+\omega_f}{1+\omega_i}\frac{5+3\omega_i}{5+3\omega_f}\Phi(\eta_i) . \tag{2.29}$$

This yields for the case of a transition from a radiation-dominated Universe with $\omega_i = \frac{1}{3}$ to a matter-dominated one with $\omega_f = 0$, that the gravitational potential drops by a factor of $9/10$:

$$\Phi(\eta_f) = \frac{9}{10}\Phi(\eta_i) \,. \tag{2.30}$$

From this equation we deduce how to relate the potential for regions far away from the transition.

For the radiation/matter system, it is possible to derive an exact solution displaying the full time dependence. Defining $\varkappa := \left(\sqrt{2}-1\right)\eta/\eta_{\text{eq}}$, one obtains (c.f. [Muk05])

$$\Phi(\varkappa) = \frac{\varkappa+1}{(\varkappa+2)^3}\left[A\left(\frac{3}{5}\varkappa^2 + 3\varkappa + \frac{1}{\varkappa+1} + \frac{13}{3}\right) + B\frac{1}{\varkappa^3}\right], \tag{2.31}$$

where A and B are integration constants.

2.5.2 Short-Wavelength Solutions

In the region where $c_s\eta k \gg 1$, one can neglect the last term on the right-hand side of (2.26). The result is (c.f. [Muk05])

$$v_k'' + c_s^2 k^2 v_k \simeq 0 \,, \tag{2.32}$$

which describes *sound waves* with time-dependent amplitude.

2.6 Baryon-Radiation Plasma and Cold Dark Matter

The ideal-fluid approximation made so far does not take into account that photons can transfer energy from regions of the fluid over distances due to their mean free path. Taking, in turn, *shear viscosity* into account, the energy momentum tensor (2.20) acquires an additional contribution proportional to the shear viscosity coefficient η_{vis} and reads (c.f. [Muk05])[1]

$$T^\mu_{\ \nu} = (\varepsilon + p)\, u^\mu u_\nu - p\,\delta^\mu_{\ \nu} - \eta_{\text{vis}}\left(P^\mu_{\ \gamma}u_\nu^{\ ;\gamma} + P^\gamma_{\ \nu}u^\mu_{\ ;\gamma} - \tfrac{2}{3}P^\mu_{\ \nu}u^\gamma_{\ ;\gamma}\right). \tag{2.33}$$

Here, $P^\mu_{\ \nu} := \delta^\mu_{\ \nu} - u^\mu u_\nu$ is a projection operator, and a semicolon denotes a covariant derivative with respect to the corresponding coordinate. One can show [Tho30] that the shear viscosity coefficient is given by $\eta_{\text{vis}} = 4/15\,\varepsilon_\gamma\,\tau_\gamma$, where τ_γ is the mean free time of the photons (c.f. also [Wei71, And76]).

---

[1] In principle, one would also have to take the *bulk viscosity* into account. However, one can show that it is negligible for the case at hand [Wei71]. There is also a general theorem, proven be Tisza [Tis42], stating that the bulk viscosity coefficient ζ_{bul} vanishes if the trace of the energy-momentum tensor is a function solely of the energy density ε and the particle number density n. In particular, $\zeta_{\text{bul}} = 0$ for a gas of structureless point particles in the extreme relativistic and non-relativistic limits. However, negligible ζ_{bul} is the exception, rather than the rule for general imperfect fluids [Tis42].

2.6 Baryon-Radiation Plasma and Cold Dark Matter

Opposed to the case without shear viscosity, the potentials Φ and Ψ are a priori *not* equal, which is only the case if the spatial part of the energy-momentum tensor vanishes, being obviously not fulfilled under the given circumstances. Nevertheless, it is easy to show that their difference, i.e. $\Phi - \Psi$, is in any case suppressed by the ratio of the mean free path of the photons to the perturbation scale, and is strongly damped after the radiation/matter equality. Hence, in the following we will set $\Phi \equiv \Psi$.

Using conformal-Newton gauge, one finds (c.f. [Muk05]) from the conservation law $T^\mu{}_{\nu;\mu} = 0$ to first order in the perturbations and for $\nu = 0$ that

$$\delta\varepsilon' + 3\mathcal{H}(\delta\varepsilon + \delta p) - 3(\varepsilon + p)\Phi' + a(\varepsilon + p)u^i{}_{,i} = 0, \tag{2.34a}$$

in which we observe the absence of shear viscosity. The case of $\nu = i$ leads to (c.f. [Muk05])

$$\frac{1}{a^4}\left(a^5(\varepsilon + p)u^i{}_{,i}\right)' - \frac{4}{3}\eta_{\text{vis}}\Delta u^i{}_{,i} + \Delta\delta p + (\varepsilon + p)\Delta\Phi = 0. \tag{2.34b}$$

These equations are separately valid for the components of the baryon-radiation plasma and the dark matter.

Using the **photon density contrast**

$$\delta_\gamma := \frac{\delta\varepsilon_\gamma}{\varepsilon_\gamma}, \tag{2.35}$$

and (2.34a) one finds

$$(\delta_\gamma - 4\Phi)' + \frac{4}{3}a u^i{}_{,i} = 0. \tag{2.36}$$

This expression allows us to express the gradient $u^i{}_{,i}$ in terms of δ_γ and S, and to write equation (2.34b) in the form (c.f. [Muk05])

$$\left(\frac{\delta'_\gamma}{c_s^2}\right)' - \frac{3\eta_{\text{vis}}}{\varepsilon_\gamma a}\Delta\delta'_\gamma - \Delta\delta_\gamma = \frac{4}{3c_s^2}\Delta\Phi + \left(\frac{4\Phi'}{c_s^2}\right)' - \frac{12\eta_{\text{vis}}}{\varepsilon_\gamma a}\Delta\Phi'. \tag{2.37}$$

2.6.1 Long-Wavelength Solutions

In the case of long-wavelength perturbations, for which one can neglect the $u^i{}_{,i}$ term, equation (2.36) implies

$$\delta_\gamma - 4\Phi = \text{const.} \tag{2.38}$$

Above, the constant on the right-hand side can be determined by noting that within the radiation-dominated area, the gravitational potential stays constant on super-horizon scales and is primarily due to fluctuations in the radiation component. Thus,

$$\delta_\gamma \simeq -2\Phi(\eta \ll \eta_{\text{eq}}) \equiv -2\Phi_0, \tag{2.39}$$

which implies with

$$\Phi(\eta \gg \eta_{\text{eq}}) \simeq \frac{9}{10} \Phi_0 \qquad (2.40)$$

at recombination (subscript 'r') the results

$$\delta_\gamma(\eta_r) \simeq -\frac{8}{3} \Phi(\eta_r) \simeq -2.4\,\Phi_0\,, \qquad (2.41\text{a})$$

$$\delta'_\gamma(\eta_r) \simeq 0\,. \qquad (2.41\text{b})$$

These formulae will be used in the following chapter when we discuss cosmic microwave background anisotropies.

2.6.2 Short-Wavelength Solutions

At the end of the following chapter we will discuss acoustic peaks in the spectrum of the cosmic microwave background radiation. Since these are generated by short-wavelength perturbations, i.e. those with $\eta_r\, k > 1$, we will discuss them next.

To get analytically tractable solutions, we assume that the speed of sound is slowly varying, neglect the contribution of the baryons to the gravitational potential, since they form just a small fraction of the total matter density, and consider only times $\eta > \eta_{\text{eq}}$, which allows us to take the gravitational potential constant in time. This last fact reduces equation (2.37) to

$$\frac{\mathrm{d}^2 \delta_\gamma}{\mathrm{d}x^2} - \frac{4\,\tau_\gamma}{5\,a}\Delta\frac{\mathrm{d}\delta_\gamma}{\mathrm{d}x} - \frac{1}{c_s^2}\Delta\delta_\gamma = \frac{4}{3\,c_s^4}\Delta\Phi\,, \qquad (2.42)$$

where we changed to the variable x, defined by $\mathrm{d}x \equiv c_s^2\,\mathrm{d}\eta$. The solution to this linear differential equation can be expressed as the sum of a general solution to the homogeneous equation (r.h.s. = 0) plus a particular solution of the inhomogeneous one. The latter is obviously given by

$$\delta_\gamma^{\text{part}}(\eta, k) \simeq -\frac{4}{3\,c_s^2(\eta)}\Phi(k)\,. \qquad (2.43\text{a})$$

Using the assumptions mentioned at the beginning of this subsection together with the WKB approximation, one finds for the former:

$$\delta_\gamma^{\text{hom}}(\eta, k) \simeq A(k)\,\sqrt{c_s(\eta)}\,\cos\!\left(k\int_0^\eta \mathrm{d}\rho\, c_s(\rho)\right) e^{-(k/k_D(\eta))^2}, \qquad (2.43\text{b})$$

where $A(k)$ is an integration constant, and the **dissipation scale** k_D is given by (c.f. [Muk05])

$$k_D(\eta) := \left(\frac{2}{5}\int_0^\eta \mathrm{d}\rho\,\frac{c_s^2(\rho)\,\tau_\gamma(\rho)}{a(\rho)}\right)^{-1/2}. \qquad (2.44)$$

Thus, the full solution of (2.42) reads

$$\delta_\gamma(\eta, k) \simeq -\frac{4}{3 c_s^2(\eta)} \Phi(k) + A(k) \sqrt{c_s(\eta)} \cos\left(k \int_0^\eta \mathrm{d}\rho\, c_s(\rho)\right) e^{-(k/k_D(\eta))^2}, \qquad (2.45)$$

It can be shown that for vanishing shear viscosity together with constant sound speed, the above equation is also valid in the regime $k\eta \ll 1$.

Eventually, we would like to have a relation between the primordial gravitational potential Φ_0 and the photon density contrast evaluated after equality. For the long-wavelength modes we found in section 2.5.1 that Φ drops by a factor of $9/10$. Similarly one can show (c.f. [Muk05]) that for short-wavelength perturbations the following equation holds

$$\Phi(\eta > \eta_{\mathrm{eq}}, k) = \frac{\ln(0.15\, k\, \eta_{\mathrm{eq}})}{(0.27\, k\, \eta_{\mathrm{eq}})^2} \Phi_0(k)\,. \qquad (2.46)$$

Furthermore, one finds (c.f. [Muk05]) that the integration constant $A(k)$ is given by

$$A\!\left(k \gg \eta_{\mathrm{eq}}^{-1}\right) \simeq 6 \sqrt[4]{3}\, \Phi_0(k)\,, \qquad (2.47)$$

whereas for perturbations which enter the horizon long after equality, the result is (c.f. [Muk05])

$$A\!\left(k \ll \eta_{\mathrm{eq}}^{-1}\right) \simeq \frac{6}{5} \sqrt[4]{3}\, \Phi_0(k) = \frac{1}{5} A\!\left(k \gg \eta_{\mathrm{eq}}^{-1}\right)\,. \qquad (2.48)$$

We will, however, postpone further discussions on this subject until section 3.4.1 and continue now with the study of the origin of primordial inhomogeneities.

2.7 Origin of Primordial Inhomogeneities

So far, we have studied gravitational instabilities in a Universe filled with hydro-dynamical matter. For a better understanding of the generation of primordial inhomogeneities, we will now extend our analysis to a study of scalar-field condensates. We will study the behaviour of perturbations during an inflationary stage and close with a discussion of gravitational waves from inflation.

Let us consider a flat Universe filled with a scalar-field condensate, described by the action

$$\mathcal{S}[X] = \int \mathrm{d}^4 x\, \sqrt{-g}\, \mathcal{L}\big(X(\varphi), \varphi\big)\,, \qquad (2.49)$$

where

$$X(\varphi) := \frac{1}{2} g^{\mu\nu} \varphi_{,\mu} \varphi_{,\nu}\,. \qquad (2.50)$$

Setting

$$u_\mu := \frac{\varphi_{,\mu}}{\sqrt{2X}}, \qquad p := \mathcal{L}, \qquad \varepsilon := 2Xp_{,X} - \mathcal{L} \qquad (2.51)$$

it is possible to write the corresponding energy-momentum tensor in an ideal-fluid form,

$$\mathrm{T}^\mu{}_\nu = (\varepsilon + p)\, u^\mu u_\nu - p\, \delta^\mu{}_\nu \, . \qquad (2.52)$$

For the canonical scalar field with $\mathcal{L} = X - \mathrm{V}(\varphi)$ one has $\varepsilon = X + \mathrm{V}(\varphi)$. In general, ε, φ and X have to be treated as independent variables and it is not possible to write down an equation of state in the form $p = p(\varepsilon)$. However, if $\mathcal{L} = X^n$ we have $p = \varepsilon/(2n-1)$, which in the case of $n = 2$ describes an ultra-relativistic fluid, where $p = \varepsilon/3$.

2.7.1 Perturbations

We will now derive equations for inhomogeneities caused by the model (2.49). For simplicity we will work in conformal-Newton gauge to express the gauge-invariant perturbations of the energy-momentum tensor in terms of metric and scalar field perturbations.

Analogous to the gauge-invariant energy-density perturbation $\delta\bar{\varepsilon}$ introduced in equation (2.21a), we define the gauge-invariant scalar-field perturbation $\delta\bar{\varphi}$ via

$$\delta\bar{\varphi} := \delta\varphi - \varphi_0'\left(\mathrm{B} - \mathrm{E}'\right), \qquad (2.53)$$

where φ_0 denotes the homogeneous field subject to the unperturbed background. Expressing the speed of sound via

$$c_s^2 = \frac{p_{,X}}{\varepsilon_{,X}}, \qquad (2.54)$$

and defining the variables u and v through

$$u := \sqrt{\varepsilon_{,X}}\, a \left(\delta\bar{\varphi} + \frac{\varphi_0'}{\mathcal{H}}\Psi\right), \qquad v := \frac{2\Psi}{\sqrt{\varepsilon + p}}, \qquad (2.55)$$

one finds

$$u'' - c_s^2 \Delta u - \frac{z''}{z} u = 0 \, , \qquad (2.56a)$$

$$v'' - c_s^2 \Delta v - \frac{\vartheta''}{\vartheta} v = 0 \, . \qquad (2.56b)$$

Here, the quantities z and ϑ are given by

$$z := \frac{1}{c_s} a\sqrt{1+\omega}\,, \qquad \vartheta := \frac{1}{c_s z} = \frac{1}{a}\frac{1}{\sqrt{1+\omega}} \, . \qquad (2.57)$$

2.7 Origin of Primordial Inhomogeneities

We note that equation (2.56b) is the same as (2.26), which is also clear from the ideal-fluid form of $T_{\mu\nu}$, c.f. (2.52). The corresponding solutions have already been discussed in section 2.5. Here, however, the meaning is different as they describe scalar-field perturbations. We briefly review the corresponding results.

If these are of *short-wavelength*, i.e. if $c_s k \gg \sqrt{|\vartheta''|/|\vartheta|}$, the WKB approximation yields [Muk05]

$$\Phi \propto \dot{\varphi}_0 \sqrt{\frac{p_{,X}}{c_s}} \exp\left(\pm i k \int dt \, \frac{c_s(t)}{a(t)}\right), \tag{2.58a}$$

$$\delta\bar{\varphi} \propto \sqrt{\frac{1}{c_s p_{,X}}} \left(\pm i c_s \frac{k}{a} + H + \ldots\right) \exp\left(\pm i k \int dt \, \frac{c_s(t)}{a(t)}\right). \tag{2.58b}$$

Thus, the scalar-field and metric perturbations *oscillate* in the short-wavelength regime. In case of *long-wavelength* perturbation, i.e. for $c_s k \ll \sqrt{|\vartheta''|/|\vartheta|}$, one obtains

$$\Phi \propto 1 - \frac{H}{a} \int dt \, a(t), \tag{2.59a}$$

$$\delta\bar{\varphi} \propto \frac{\dot{\varphi}_0}{a} \int dt \, a(t). \tag{2.59b}$$

2.7.2 Canonical Scalar Field

Having formally discussed a general class of scalar-field models described by the action (2.49), we now work out explicitly the example of a minimally-coupled canonical scalar field in a potential.

In the following paragraph, we will use conformal-Newton gauge to perform explicit calculations and start with the action (1.17), i.e.

$$S[\varphi] = \int d^4x \sqrt{-g} \, \mathcal{L}(\varphi) = \int d^4x \sqrt{-g} \left[\frac{1}{2} g^{\mu\nu} \varphi_{,\mu} \varphi_{,\nu} - V(\varphi)\right]. \tag{2.60}$$

Then one finds with equation (2.19b) if $\epsilon_{sr} \ll 1$ (in cosmic time t),

$$\dot{\Phi} + H\Phi = \frac{1}{2} \dot{\varphi}_0 \, \delta\varphi = \epsilon_{sr} H^2 \frac{\delta\varphi}{\dot{\varphi}_0}, \tag{2.61}$$

on super-horizon scales for the **comoving curvature perturbation**

$$\mathcal{R} := \psi + H \frac{\delta\varphi}{\dot{\varphi}_0} \simeq H \frac{\delta\varphi}{\dot{\varphi}_0}. \tag{2.62}$$

By definition, this quantity is gauge-invariant.

Thus we observe that the spatial behaviour of $\mathcal{R}$ is directly determined by that of $\delta\varphi$. The knowledge of $\delta\varphi$ is therefore crucial in determining correlation functions of $\mathcal{R}$ like the dimensionless **curvature power spectrum**

$$\mathcal{P}_{\mathcal{R}}(k) := k^3 \langle \mathcal{R}_k \mathcal{R}_k^* \rangle, \tag{2.63}$$

where, the angle brackets represent quantum averages. Concerning the k-dependence, one has from (2.62) the relation

$$\mathcal{P}_{\mathcal{R}}(k) \propto \mathcal{P}_{\varphi}(k) := k^3 \langle \delta\varphi_k \delta\varphi_k^* \rangle. \tag{2.64}$$

In chapter 6 we show that for free minimally-coupled scalar fields with mass μ, the result on superhorizon scales reads

$$\mathcal{P}_{\varphi}(k) \simeq k^{3-\nu}, \tag{2.65}$$

with

$$\nu \simeq 3 - \frac{2}{3}\frac{\mu^2}{H^2} + 2\epsilon_{sr} = 3 - 2\eta_{sr} + 2\epsilon_{sr}. \tag{2.66}$$

Hence, up to corrections in $\eta_{sr} \ll 1$ and $\epsilon_{sr} \ll 1$ one finds that the spectra are **scale invariant**, i.e. $\mathcal{P}$ is constant in k.

As we have seen in chapter 1, generic single-field inflation models need a flat potential, which in particular means that the mass has to be small. According to this we arrive at the conclusion that single-field inflation *predicts* an almost scale-invariant curvature spectrum. Another crucial prediction is the generation of gravitational waves, which we discuss subsequently.

2.7.3 Inflationary Gravitational Waves

Similar to the previous subsection, the study of gravitational waves starts with the corresponding action, which reads in this case to second order in the tensor h:

$$\mathcal{S}[\mathrm{h}] = \frac{1}{8}\int \mathrm{d}\eta\,\mathrm{d}^3x\, a(\eta)^2\, \mathrm{Tr}\Big[\big(\mathrm{h}(\eta,\boldsymbol{x})'\big)^2 - \big(\nabla \mathrm{h}(\eta,\boldsymbol{x})\big)^2\Big]. \tag{2.67}$$

After substituting the expansion

$$\mathrm{h}_{ij}(\eta,\boldsymbol{x}) = \int \frac{\mathrm{d}^3 k}{(2\pi)^3}\, \mathrm{h}(\eta,\boldsymbol{k})\, \mathrm{e}_{ij}(\boldsymbol{k})\, \mathrm{e}^{-\mathrm{i}\boldsymbol{k}\cdot\boldsymbol{x}}, \tag{2.68}$$

with the polarisation tensor $\mathrm{e}(\boldsymbol{k})$ and the definition

$$\mathrm{v}(\eta,\boldsymbol{k}) := \sqrt{\mathrm{Tr}[\mathrm{e}(\boldsymbol{k})^2]}\, a(\eta)\, \mathrm{h}(\eta,\boldsymbol{k}), \tag{2.69}$$

2.7 Origin of Primordial Inhomogeneities

one finds that (2.67) in a de Sitter Universe reduces to

$$S[h] = \frac{1}{2}\int d\eta\, d^3k \left[|v(\eta, \boldsymbol{k})'|^2 - \left(k^2 - \frac{2}{\eta^2}\right)|v(\eta, \boldsymbol{k})|^2\right]. \tag{2.70}$$

This is just the equation for a free, massless scalar field discussed in chapter 6, which implies a scale-invariant spectrum on super-horizon scales. Furthermore, it can be shown (c.f. [Muk05]) that in terms of the physical wavelength $\lambda_{\rm ph}$, the dimensionless power spectrum for primordial gravitational waves obeys

$$\mathcal{P}_{\rm h}(\lambda_{\rm ph}) \sim \begin{cases} \lambda_{\rm ph} & : \quad \lambda_{\rm ph} < H_0^{-1} z_{\rm eq}^{-1/2}\,, \\ \lambda_{\rm ph}^2 & : \quad H_0^{-1} z_{\rm eq}^{-1/2} < \lambda_{\rm ph} < H_0^{-1}\,, \\ {\rm const.} & : \quad H_0^{-1} < \lambda_{\rm ph}\,, \end{cases} \tag{2.71}$$

where its amplitude on scales of a few light years is about 10^{-17} for realistic models of inflation.

Chapter 3
Cosmic Microwave Background Anisotropies

»*Les miracles véritables, qu'ils font peu de bruit!*«

ANTOINE DE SAINT-EXUPERY

The **cosmic microwave background** (CMB) radiation consists of photons that —after recombination — freely stream through the Universe, basically without further scattering. It provides a snapshot of the radiation distribution at redshift $z \approx 1000$ when photons last interacted with matter.

As predicted by inflation, this radiation is very homogeneous — relative fluctuations are of order 10^{-5} on average. It is interesting that this is just the amount needed to explain the formation of large-scale structures in the case of a Universe with cold, dark and ordinary matter.

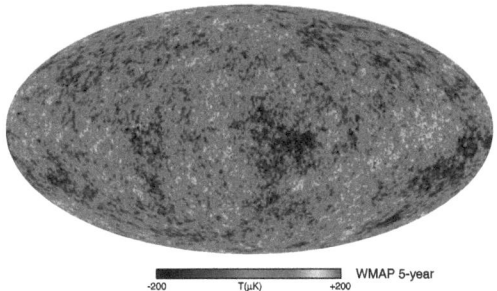

Figure 3.1: A foreground-reduced map (internal linear combination) based on the five-year WMAP data. (Figure taken from [LAMDA].)

In figure 3.1 we show the CMB fluctuations across the entire sky, as measured by the WMAP satellite and processed by the WMAP collaboration [LAMDA]. On this map, an angular diameter of $\theta \approx 1°$ corresponds to the Hubble radius at recombination, dividing small- and large-scale inhomogeneities. While the former have entered the horizon before recombination, hence being vitally influenced by gravitational instabilities, the latter have not changed since inflation and thus provide a direct insight into the primordial spectra of perturbations.

This chapter deals with a derivation of the dominant physical effects inducing fluctuations in the CMB radiation. The discussion shall be restricted to the leading large-scale effects. On the one hand this sacrifices some accuracy but has, on the other hand, the advantage of being analytically tractable, providing hence solid and clear analytical insights in all major dependencies. In our discussion we mainly follow, and in part summarise, [Muk05].

3.1 Boltzmann Equation and Temperature Fluctuations

In this section we first discuss the Boltzmann equation, which is then, in section 3.2, used to derive a relation between the temperature fluctuations and the gravitational potential. We also focus on the temperature spectrum and the coordinate transformation properties of the CMB temperature.

Given an ensemble of identical particles, it follows from the invariance of the **phase-space volume**

$$d^3x \, d^3p := dx^1 \, dx^2 \, dx^3 \, dp_1 \, dp_2 \, dp_3 \tag{3.1}$$

under general coordinate transformations,

$$x \to \tilde{x}(x) \,, \tag{3.2}$$

that the one particle **phase-space distribution** f, defined via

$$dN \equiv f(\boldsymbol{x}, \boldsymbol{p}, \eta) \, d^3x \, d^3p \,, \tag{3.3}$$

is a space-time scalar. Here, dN is the number of particles with the volume $d^3x \, d^3p$ and, as usual, a symbol in bold face denotes a 3-vector with *contravariant* components.

If these particles are stable and non-interacting, the particle number does not change with time within this volume element, and hence, the distribution function obeys the collisionless **Boltzmann equation**

$$\frac{d f(\boldsymbol{x}, \boldsymbol{p}, \eta)}{d \eta} = \frac{\partial f(\boldsymbol{x}, \boldsymbol{p}, \eta)}{\partial \eta} + \frac{d\boldsymbol{x}}{d\eta} \cdot \nabla_{\boldsymbol{x}} f(\boldsymbol{x}, \boldsymbol{p}, \eta) + \frac{d\boldsymbol{p}}{d\eta} \cdot \nabla_{\boldsymbol{p}} f(\boldsymbol{x}, \boldsymbol{p}, \eta) = 0 \,. \tag{3.4}$$

3.1 Boltzmann Equation and Temperature Fluctuations

It is experimentally well established that the distribution $f(\boldsymbol{x}, \boldsymbol{p}, \eta)$ of the CMB radiation is Planckian to good precision and hence given by

$$f(\boldsymbol{x}, \boldsymbol{p}, \eta) \equiv f(\omega/T) := \frac{2}{\exp(\omega/T) - 1} \,. \tag{3.5}$$

Additionally, it is known that the CMB temperature T is extremely isotropic. Therefore,

$$T(\eta, \boldsymbol{x}, \boldsymbol{l}) = T_0(\eta) + \delta T(\eta, \boldsymbol{x}, \boldsymbol{l}) \,, \tag{3.6}$$

where the temperature fluctuations δT are small compared to the CMB background temperature T_0, i.e. $\delta T \ll T_0$, and η might take any value between recombination and today. Above, $-\boldsymbol{l} := \boldsymbol{p}/\|\boldsymbol{p}\|$ is the direction of the photon momentum. For any observer with four-velocity u^α, the frequency ω of the radiation can be expressed as

$$\omega = p_\alpha u^\alpha \,. \tag{3.7}$$

Now, in the rest frame of each observer, we have

$$g_{00} \left(u^0 \right)^2 = 1 \tag{3.8}$$

and hence the *physical* frequency ω_{ph} is given by

$$\omega_{\text{ph}} = \omega = \frac{p_0}{\sqrt{g_{00}}} \,. \tag{3.9}$$

Using the latter formula and $p_\alpha p^\alpha = 0$ one can relate the frequency ω measured by an observer $\mathcal{O}$, to the frequency $\tilde{\omega}$ measured by another observer $\tilde{\mathcal{O}}$, who's frame is related to $\mathcal{O}$'s via

$$\tilde{x} = x + \zeta \,, \tag{3.10}$$

for all $i, j = 1, 2, 3$. Let us now calculate to first order in ζ and its derivatives. Equation (3.10) then implies

$$\tilde{g}_{00} = \frac{\partial x^\alpha}{\partial \tilde{\eta}} \frac{\partial x^\beta}{\partial \tilde{\eta}} g_{\alpha\beta} \simeq g_{00} - 2 \frac{\partial \zeta^0}{\partial \tilde{\eta}} \,, \tag{3.11}$$

and

$$\tilde{p}_0 = \frac{\partial x^\beta}{\partial \tilde{\eta}} p_\beta \simeq p_0 - \frac{\partial \zeta^\beta}{\partial \tilde{\eta}} p_\beta \,, \tag{3.12}$$

from which we obtain

$$\tilde{\omega} \simeq \omega \left(1 + \frac{\partial \zeta}{\partial \eta} \cdot \boldsymbol{l} \right) . \tag{3.13}$$

Since f is a Lorentz-scalar, ω/T is too and thus from

$$\tilde{T} \simeq T_0 - T_0' \zeta^0 \qquad (3.14)$$

we have that the temperature fluctuations in $\tilde{\mathcal{O}}$ are related to those in $\mathcal{O}$ via

$$\delta\tilde{T} \simeq \delta T - T_0' \zeta^0 + T_0 \frac{\partial \zeta}{\partial \eta} \cdot \boldsymbol{l}\,. \qquad (3.15)$$

The **monopole** (l-independent term) and the **dipole** (term proportional to l) depend on the particular rest frame of the observer. Because up to now, we can only measure the CMB radiation accurately from one vantage point, the monopole can be removed by a redefinition of the background temperature. The dipole relies on the observer's motion relative to the frame of the background radiation. Thus we will subtract them in the following part of this thesis since we are interested in information inherent to the initial temperature fluctuations.

3.2 Sachs-Wolfe Effect

With the Boltzmann equation and the transformation properties of the CMB tempera-ture — derived and discussed in the previous section — together with the results from chapter 2, we have all necessary tools at hand to study the spectrum of temperature fluctuations. Since our goal will be to calculate temperature correlations on scales that exceed the horizon at recombination, $\theta_r \approx 0.87°$ in a flat Universe, we focus on large-scale effects only.

Assuming instantaneous decoupling, we may impose a matching condition of a hydrodynamical energy-momentum tensor (describing the radiation before decoupling) and its kinetic pendant (immediately afterwards). For the 0-0-component, this means

$$\varepsilon_\gamma(1+\delta_\gamma) = \mathrm{T^{(hyd)}}{}^0{}_0 \stackrel{!}{=} \mathrm{T^{(kin)}}{}^0{}_0 = \frac{1}{\sqrt{-g}} \int d^3p\, \mathrm{f}\!\left(\frac{\omega}{T}(\boldsymbol{p})\right) p_0\,. \qquad (3.16)$$

To find a relation between the density contrast $\delta_\gamma := \delta\varepsilon_\gamma/\varepsilon_\gamma$ of the radiation and the relative temperature fluctuations, we now solve the Boltzmann equation for freely propagating radiation. This shall be done in an almost flat Universe using the conformal-Newtonian gauge, in which the metric takes the form (neglecting anisotropic stress)

$$g_{\mu\nu} = a^2\left(\eta_{\mu\nu} + 2\,\Phi\,\delta_{\mu\nu}\right). \qquad (3.17)$$

Here, the metric potential Φ of the scalar field fluctuations is assumed to be small, i.e. $\Phi \ll 1$. In this metric, it follows from $p_\alpha p^\alpha = 0$ that

$$p^0 = \frac{\sqrt{\|\boldsymbol{p}^2\|}}{a^2} \equiv \frac{p}{a^2} \qquad (3.18\mathrm{a})$$

and

$$p_0 = (1 + 2\Phi) p \,. \tag{3.18b}$$

Taking into account the geodesic equation for radiation

$$\frac{\mathrm{d}p_\alpha}{\mathrm{d}\lambda} = \frac{1}{2} \frac{\partial g_{\gamma\delta}}{\partial x^\alpha} p^\gamma p^\delta \,, \qquad \frac{\mathrm{d}x^\alpha}{\mathrm{d}\lambda} = p^\alpha \,, \tag{3.19}$$

where λ is an affine parameter along the geodesic, we find

$$\frac{\mathrm{d}\boldsymbol{x}}{\mathrm{d}\eta} = \frac{\boldsymbol{p}}{p^0} = (1 + 2\Phi) \boldsymbol{l} \tag{3.20}$$

and hence, with the metric (3.17),

$$\frac{\mathrm{d}p_\alpha}{\mathrm{d}\eta} \simeq 2p \frac{\partial \Phi}{\partial x^\alpha} \,. \tag{3.21}$$

According to this, the Boltzmann equation can be written as

$$\frac{\partial \mathrm{f}}{\partial \eta} = -(1 - 2\Phi) \boldsymbol{l} \cdot \nabla_{\boldsymbol{p}} \mathrm{f} - 2p (\nabla_{\boldsymbol{x}} \Phi) \cdot (\nabla_{\boldsymbol{p}} \mathrm{f}) \,, \tag{3.22}$$

which shows for the distribution (3.5) to *zeroth order* in Φ and $\delta T/T$ that

$$(T_0 \, a)' = 0 \,, \tag{3.23}$$

therefore giving that the radiation temperature is inversely proportional to the scale factor. The *first order* of (3.22) reads

$$\left(\frac{\partial}{\partial \eta} + \boldsymbol{l} \cdot \nabla_{\boldsymbol{x}} \right) \left(\frac{\delta T}{T} + \Phi \right) = 2 \frac{\partial \Phi}{\partial \eta} \,, \tag{3.24}$$

where we used the fact that

$$\frac{\omega}{T} = \frac{p_0}{T \sqrt{g_{00}}} \simeq \frac{p}{T_0 a} \left(1 + \Phi - \frac{\delta T}{T} \right) . \tag{3.25}$$

Equation (3.24) determines the temperature fluctuations of the CMB.

When we apply the results of this chapter to the main results of this thesis in chapter 6, we will restrict ourselves to the case of a constant gravitational potential, i.e. $\frac{\partial \Phi}{\partial \eta} = 0$. Along null geodesics we then find the obvious solution to equation (3.24)

$$\frac{\delta T}{T} + \Phi = \text{const.} \tag{3.26}$$

This result, among others, was first derived by Sachs and Wolfe [SW67] and is therefore called **ordinary Sachs-Wolfe effect**. If Φ depends on time, e.g. as a result of a small fraction of radiation after recombination or if some form of dark energy is present, we speak about the **integrated Sachs-Wolfe effect**. As numerical calculations show, the effect does practically never contribute more than 10–15 % for an almost scale-invariant spectrum (c.f. [Muk05]).

Using (3.18b) and (3.25), the right hand side of (3.16) might be expressed as

$$\frac{1}{\sqrt{-g}} \int d^3p \, f\!\left(\frac{\omega}{T}(\boldsymbol{p})\right) p_0 \simeq \frac{1}{a^4(1-2\Phi)} \int d^3p \, f\!\left(\frac{p_0}{\sqrt{g_{00}}\,T}(\boldsymbol{p})\right) p_0$$

$$\simeq \int d^3p \, f\!\left(\frac{p_0}{aT_0}\left[1+\Phi-\frac{\delta T}{T}\right](\boldsymbol{p})\right) \frac{(1-4\Phi)p}{a^4}$$

$$\simeq \left(4\pi T_0^4 \int dy \, f(y) \, y\right) \int \frac{d^2l}{4\pi}\left(1+4\frac{\delta T}{T}(\boldsymbol{l})\right) \quad (3.27)$$

$$\equiv \varepsilon_\gamma \left(1+4 \int \frac{d^2l}{4\pi}\frac{\delta T}{T}(\boldsymbol{l})\right)$$

$$\stackrel{!}{=} \varepsilon_\gamma (1+\delta_\gamma)\,,$$

which implies

$$\delta_\gamma = 4 \int \frac{d^2l}{4\pi}\frac{\delta T}{T}(\boldsymbol{l})\,. \quad (3.28)$$

As has been shown in the previous chapter, the gravitational potential Φ and the density contrast of the radiation at recombination (subscript 'r') can, for $k\eta_r \ll 1$, be related as

$$\delta_\gamma(k,\eta_r) \simeq -\frac{3}{8}\Phi(k,\eta_r)\,, \quad (3.29a)$$

$$\delta'_\gamma(k,\eta_r) \simeq 0\,. \quad (3.29b)$$

From equation (3.26) we have for the temperature fluctuation today (subscript '0')

$$\frac{\delta T}{T}(\eta_0,\boldsymbol{x}_0,\boldsymbol{l}) = \frac{\delta T}{T}(\eta_r,\boldsymbol{x}_r,\boldsymbol{l}) + \Phi(\eta_r,\boldsymbol{x}_r) - \Phi(\eta_0,\boldsymbol{x}_0)\,, \quad (3.30)$$

and hence, by virtue of (3.28) together with (3.29a,b) one obtains

$$\frac{\delta T}{T}(\eta_0,\boldsymbol{x}_0,\boldsymbol{l}) \simeq \frac{1}{3}\Phi(\eta_r,\boldsymbol{x}_0-\boldsymbol{l}\eta_0)\,, \quad (3.31)$$

where we subtracted the gravitational potential today, since it is l-independent and thus just contributes to the monopole. Thus, the CMB temperature fluctuations *today* are given by one-third of the gravitational potential at the time of *recombination*.

3.3 Temperature Correlation

Having derived an approximate large-scale expression for the temperature fluctuations of the cosmic microwave background in the last section, we will now study their spectrum.

So far, precision experiments today measure only the CMB radiation on or nearby the earth, which on cosmological scales count like a point measurement. Under the *assumption* of homogeneity and isotropy, it should be approximately equivalent to take a patch of certain size of the CMB sky and to average over all possible *different patches* from a single vantage point, than averaging the data from given directions at *different places*, being more than one Hubble distance apart from each other. The smaller the observed region in question, the more of such patches are found, the better the statistics are, and the smaller the difference between the two averaging procedures is. The root-mean square of this difference is called **cosmic variance**. Another uncertainty is caused by the fact that often only a part of the sky is used for the analysis and the associated difference to the full-sky study is termed **sample difference**. Of course, to those uncertainties one also has to add uncertainties from the measurement instrument itself.

Because there are other sources than the CMB that radiate in the observed frequency bands of ongoing and future CMB experiments, one tries to reduce this contamination and in turn cuts out a region of the sky associated with our own galaxy, which provides the most dominant contribution of extrinsic radiation to the CMB (c.f. figure 3.2). Depending on the specific exclusion mask one has to adjust the calculation and interpretation of the results inferred from temperature-fluctuation correlation functions.

In general, the spectrum of temperature fluctuations can be described by an infinite set of correlation functions

$$\mathcal{C}(\theta_{12}, \theta_{13}, \ldots, \theta_{n-1\,n}) := \left\langle \frac{\delta T}{T}(l_1) \frac{\delta T}{T}(l_2) \cdots \frac{\delta T}{T}(l_n) \right\rangle_{\theta_{12}, \theta_{13}, \ldots, \theta_{n-1\,n}} \tag{3.32}$$

with $n \in \mathbb{N}$ and the angle brackets denote an average over all possible directions l_i, for a given configuration of angles $\theta_{ik} := \arccos(l_i \cdot l_k)$. If it is nearly Gaussian, it can be described alone by the two-point function

$$\mathcal{C}(\theta) := \mathcal{C}(\theta_{12}) = \left\langle \frac{\delta T}{T}(l_1) \frac{\delta T}{T}(l_2) \right\rangle_{\theta_{12}}. \tag{3.33}$$

It is convenient to decompose this function as

$$\mathcal{C}(\theta) \equiv \sum_{\ell=2}^{\infty} \frac{2\ell+1}{4\pi} C_\ell \, P_\ell(\cos(\theta)), \tag{3.34}$$

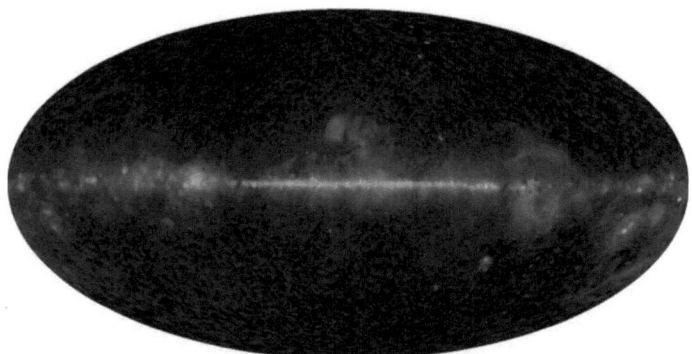

Figure 3.2: Three-color map from the maximum entropy method model for the WMAP five-year W frequency band (83.75–104.25 GHz). This map indicates which emission mechanism dominates as a function of frequency and sky position (synchrotron in red, free-free in green, thermal dust in blue). (Figure taken from [LAMDA].)

where the **multipole moments** $C_\ell \in \mathbb{R}$ are expansion coefficients, $\mathrm{P}_\ell(\cos(\theta))$ is a Legendre polynominal of degree ℓ, and we subtracted the monopole and the dipole. Again under the assumption of homogeneity and isotropy, the average over directions — appearing in the definition of $\mathcal{C}(\theta)$ — is equivalent to the average over different points, keeping the directions l_1 and l_2 fixed with the condition $\theta = \arccos(l_1 \cdot l_2)$, for a given angle θ. Hence, from the Sachs-Wolfe formula (3.31), and with $x(\eta) = x(\eta_0) + l\,(\eta - \eta_0)$ and $\eta_r/\eta_0 \simeq 0$, we have

$$\begin{aligned}
\mathcal{C}(\theta) &= \left\langle \frac{\delta T}{T}(\eta_0, \boldsymbol{x}_0, \boldsymbol{l}_1)\, \frac{\delta T}{T}(\eta_0, \boldsymbol{x}_0, \boldsymbol{l}_2) \right\rangle_\theta \\
&= \frac{1}{9} \left\langle \Phi\bigl(\eta_r, \boldsymbol{x}_0 + \boldsymbol{l}_1(\eta_r - \eta_0)\bigr)\, \Phi\bigl(\eta_r, \boldsymbol{x}_0 + \boldsymbol{l}_2(\eta_r - \eta_0)\bigr) \right\rangle_\theta \\
&\simeq \frac{1}{9} \int \mathrm{d}^3 x_0 \int \frac{\mathrm{d}^3 k}{(2\pi)^3} \int \frac{\mathrm{d}^3 p}{(2\pi)^3}\, \Phi(\eta_r, \boldsymbol{k})\, \Phi(\eta_r, \boldsymbol{p})\, \mathrm{e}^{-\mathrm{i}(\boldsymbol{k}+\boldsymbol{p})\cdot \boldsymbol{x}_0 - \mathrm{i}(\boldsymbol{k}\cdot \boldsymbol{l}_1 + \boldsymbol{p}\cdot \boldsymbol{l}_2)\eta_0} \\
&\simeq \frac{1}{9} \int \frac{\mathrm{d}^3 k}{(2\pi)^3}\, \Phi(\eta_r, \boldsymbol{k})\, \Phi^*(\eta_r, \boldsymbol{k})\, \mathrm{e}^{-\mathrm{i}\boldsymbol{k}\cdot(\boldsymbol{l}_1 - \boldsymbol{l}_2)\eta_0} \\
&= \frac{1}{2\pi^2} \int \mathrm{d}k\, k^2\, \frac{\sin\bigl(k|\boldsymbol{l}_1 - \boldsymbol{l}_2|\eta_0\bigr)}{k|\boldsymbol{l}_1 - \boldsymbol{l}_2|\eta_0}\, \frac{1}{9} |\Phi(\eta_r, k)|^2 \\
&= \frac{1}{4\pi} \int \mathrm{d}k\, k^2 \sum_{\ell=2}^{\infty} (2\ell + 1)\, [\mathrm{j}_\ell(k\eta_0)]^2\, \mathrm{P}_\ell(\cos(\theta))\, \frac{2}{9\pi} |\Phi(\eta_r, k)|^2
\end{aligned} \qquad (3.35)$$

$$\stackrel{!}{=} \sum_{\ell=2}^{\infty} \frac{2\ell+1}{4\pi} C_\ell P_\ell(\cos(\theta)),$$

where j_ℓ is a spherical Bessel function of order ℓ, implying

$$C_\ell \simeq \frac{2}{9\pi} \int dk \, k^2 \, |\Phi(\eta_r, \boldsymbol{k})|^2 \, [j_\ell(k\eta_0)]^2. \tag{3.36}$$

Taking into account that the gravitational potential on super-horizon scales drops by a factor of $\frac{9}{10}$ after radiation/matter equality and writing

$$|\Phi(\eta_r, \boldsymbol{k})|^2 \equiv \frac{9}{10} B \, k^{n_s - 4}, \tag{3.37}$$

we find by virtue of the identity

$$\int dk \, k^{m-1} [j_\ell(k)]^2 = \frac{2^{m-3} \pi \, \Gamma(2-m) \, \Gamma\!\left(\ell + \frac{m}{2}\right)}{\Gamma^2\!\left(\frac{3-m}{2}\right) \Gamma\!\left(\ell + 2 - \frac{m}{2}\right)}, \tag{3.38}$$

the result

$$C_\ell \simeq \frac{2^{n_s - 4} B}{5} \frac{\Gamma(3 - n_s) \, \Gamma\!\left(\ell + \frac{n_s - 1}{2}\right)}{\Gamma^2\!\left(2 - \frac{n_s}{2}\right) \Gamma\!\left(\ell - \frac{n_s - 5}{2}\right)}. \tag{3.39}$$

Observations [KDN+09] suggest a nearly Gaussian and **scale-invariant** power spectrum, i.e. $n_s \simeq 1$. Hence we find that the **angular power spectrum**

$$\frac{\ell(\ell+1) C_\ell}{2\pi} \tag{3.40}$$

is constant in ℓ and equal to $B/(20\pi^2)$. This flatness at low multipoles is often called **Sachs-Wolfe plateau**.

3.4 Other Effects

Although it is not necessary for the specific calculation presented in this work, we now give a concise overview over some other effects that influence the cosmic microwave background radiation. The idea behind this is to provide a more — though not fully — complete picture of the underlying physics.

3.4.1 Small-Scale Anisotropies

Perturbations on small scales, i.e. on scales that are smaller than the sound horizon at recombination, entered this horizon before recombination and hence undergo evolution, which might significantly modify the primordial spectrum. This, in turn, should also modify the angular power spectrum for multipoles higher than around $\ell_r \approx \pi/\theta \approx 200$.

The general calculation for all multipoles is quite involved and impossible to perform analytically. Hence, we will impose several assumptions and approximations. One important will be that the speed of sound changes slowly, which allows us to use the WKB subhorizon solution (2.45) after radiation/matter equality. It implies, by virtue of (3.28), (c.f. [Muk05])

$$\Phi(\eta_r, k) + \frac{\delta T}{T}(\eta_r, k) = \Phi(\eta_r, k) + \frac{1}{4}\delta_\gamma(\eta_r, k)$$
$$= T_p \left(1 - \frac{1}{3c_s^2}\right) \Phi_0(k) \qquad (3.41)$$
$$+ T_o \sqrt{c_s} \, \cos\left(k\eta_0 \varrho(\eta_r)\right) e^{-(k/k_D)^2} \Phi_0(k) ,$$

where we defined

$$\varrho := \frac{1}{\eta_0} \int_0^{\eta_r} \mathrm{d}\eta \, c_s(\eta) . \qquad (3.42)$$

Equation (3.41) also defines the **transfer functions** T_p and T_o, whose values follow from the results of section 2.6.2. They depend strongly on whether perturbations entered the horizon before or after radiation/matter equality. While for the concordance model, the former case shows that (c.f. [Muk05])

$$T_p \simeq \frac{9}{10} , \qquad T_o \simeq 0.4 , \qquad (3.43a)$$

one finds for $k\eta_{eq} \gg 1$ the result (c.f. [Muk05])

$$T_p \simeq 0 , \qquad T_o \simeq 1.97 . \qquad (3.43b)$$

The above changes have a clear physical explanation: While the gravitational potential decays for subhorizon modes during the radiation-dominated stage, it stays constant for perturbations that enter the horizon long after equality, when cold matter already dominates. Therefore, T_p changes from almost unity to zero. Since T_o defines the ampli-tude of the sound wave, its increase is caused by the gravitational field of radiation, which becomes significant when modes enter the horizon.

Next, we would like to derive an analytical expression for the multipole moments. Therefore, we note that all of the previous discussions in this chapter assume instantaneous recombination — an assumption which is certainly spoiled on small scales (**finite thickness effect**). The inherent uncertainty

3.4 Other Effects

in the time of decoupling adds a suppression on small scales and is approximately accommodated by the overall factor

$$\exp\left(-2\left(\sigma k \eta_r\right)^2\right) \qquad (3.44)$$

in the integrand of C_ℓ, equation (3.36). Therein, the **damping scale** σ depends (among others) on the ionisation fraction and (weakly) on the amount of cold dark matter and the number of light neutrinos. For the concordance model, σ is of order 10^{-2}. Taking (3.41) and the extra damping factor (3.44) to calculate the multipole moments,

$$C_\ell \simeq \frac{2}{\pi}\int dk \left|\Phi(\eta_r,k) + \frac{1}{4}\delta_\gamma(\eta_r,k)\right|^2 k^2\, e^{-2(\sigma k \eta_r)^2}\left[j_\ell(k\eta_0)\right]^2, \qquad (3.45)$$

schematically yields (c.f. [Muk05])

$$\ell(\ell+1)\, C_\ell \simeq O_\ell + N_\ell. \qquad (3.46)$$

Here, O stands for the *oscillating* and N for the *non-oscillating* contribution.

After some algebra, one obtains for the ℓ-dependence of the former (c.f. [Muk05]),

$$O_\ell \simeq \sqrt{\frac{1}{\ell}}\left[o_1\, e^{-o_2 \ell^2}\cos\left(\ell \varrho + \pi/4\right) + o_3 \cos\left(2\ell \varrho + \pi/4\right)\right]e^{-o_4 \ell^2}, \qquad (3.47)$$

where $o_1, \ldots, o_4$ are positive constants that depend on the densities Ω_m, Ω_Λ, Ω_b, the equation of state parameter w, and the Hubble constant today H_0. The oscillations due to this formula reproduce those obtained with full numerical calculations to a few percent accuracy. As can be seen in figure 3.3, the theory agrees quite well with the measurements. The ℓ-dependence of the non-oscillating part is (c.f. [Muk05])

$$N_\ell \simeq \frac{\left(n_1 - n_2 \ell^{0.3} - n_3\right)^2}{1 + n_4 \ell^{1.4}} e^{-n_5 \ell^2} + \frac{\left(n_1 - n_6 \ell^{0.3} + n_7\right)^2}{1 + n_8 \ell^{1.4}} e^{-n_9 \ell^2} + \frac{\left(n_1 - n_{10}\ell^{0.55} + n_{11}\right)^2}{1 + n_{12}\ell^2} e^{-n_9 \ell^2}, \qquad (3.48)$$

with the positive parameters $n_1, \ldots, n_{12}$ depending on Ω_m, Ω_Λ, Ω_b, w, and H_0.

The mentioned parameter dependencies of the angular power spectrum allow to determine these parameters by comparison to results of CMB measurements. The location of the first peak, for example, is basically determined by ϱ, as follows from (3.47). For the n'th extremum one finds from the first term in (3.47)

$$\ell_n = \frac{\pi}{\varrho}\left(n - \frac{1}{8}\right), \qquad (3.49)$$

which gives for the concordance model the peak values $\ell_1 \simeq 225\text{--}265$, $\ell_3 \simeq 825\text{--}865$ (c.f. [Muk05]). From the second term in (3.47), one infers that the second peak should be located at $\ell_2 \simeq 525\text{--}565$ (c.f. [Muk05]).

Additionally to the peak *locations*, one can use their *height* to determine cosmological parameters. In this manner, one arrives at the conclusion that, e.g. the height of the first peak plus the existence of the second one already implies the existence of cold dark matter, with its density exceeding that of baryons, and being less than the critical density. These statements are valid for a scale-invariant spectrum.

The cases of e.g. a spectral tilt or some from of quintessential dark matter, weaken to some extent the conclusions mentioned so far, because they increase the parameter space. Therefore, it is important to study more than just the first few peaks to constrain not only *parameters* but also cosmological *models* itself.

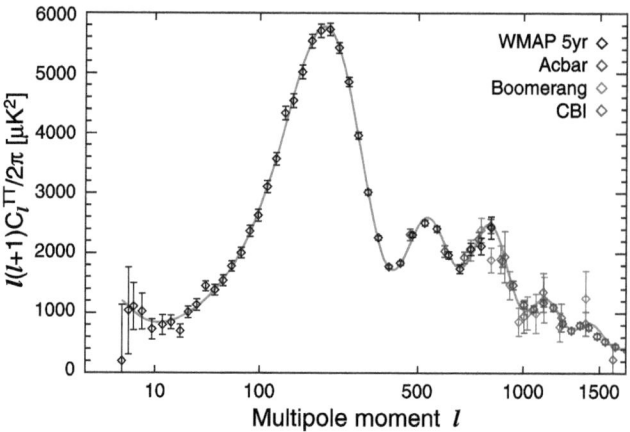

Figure 3.3: Results for the angular power spectrum, derived from the WMAP five-year, ACBAR, Boomerang, and CBI experiments. The red curve shows the best-fit ΛCDM model. (Figure taken from [NDH+09].)

3.4.2 Reionisation

Observations suggest that at redshifts of about $z \simeq 5$ most of the intergalactic hydrogen is ionised. This influences the CMB radiation, because photons can scatter on the free electrons and therefore leaves imprints in the angular power spectrum.

3.4 Other Effects

Assuming Thomson scattering for simplicity, the probability of a CMB photon to avoid interactions and to propagate freely equals (c.f. [Muk05])

$$\exp(-\tau(z)) = \exp\left(-\sigma_T \int_0^z dz' \frac{X(z')n_t}{H(z')(1+z')}\right), \quad (3.50)$$

where the **optical depth** τ depends (among others) on the redshift z, the ionisation fraction $X(z)$, the Thomson cross section σ_T and the total number of electrons n_t. Taking $H(z) \simeq H_0 \sqrt{\Omega_m}\,(1+z)^{3/2}$ and assuming $X(z < z_{\rm ion}) \simeq 1$, one finds for the concordance model

$$\tau(z_{\rm ion}) \simeq 2 \times 10^{-3}\, z_{\rm ion}^{3/2}\,. \quad (3.51)$$

The photons that rescatter change direction and may have appeared from any point remote from the original scattering point within their horizon. The multipole $\ell_{\rm ion}$ associated with the angular diameter of this horizon patch is given by

$$\ell_{\rm ion} \simeq \frac{\pi}{\Delta\theta_{\rm ion}} = \pi \frac{\eta_0}{\eta_{\rm ion}} \simeq \pi \sqrt{z_{\rm ion}}\, \Omega_m^{0.09}\,, \quad (3.52)$$

which is about $\ell_{\rm ion} \simeq 12$ for the concordance model. For multipoles larger than this value one should have an *extra damping* due to (3.50). Hence, one find for $\ell \gg \ell_{\rm ion}$

$$C_\ell^{\rm obs} = e^{-\tau}\, C_\ell\,. \quad (3.53)$$

As we shall discuss in subsection 3.4.4, there is not only a damping on small, but also a boost of the angular power spectrum on large scales around $\ell \approx \ell_{\rm ion}$.

3.4.3 Gravitational Waves

The generation of primordial **gravitational waves** is a basic prediction of inflationary cosmology. They are described by *transverse* and *traceless* metric perturbations. We therefore use the metric

$$ds^2 = a^2 \left(d\eta^2 - d\boldsymbol{x}^t(\mathbb{1} + \mathrm{h})\,d\boldsymbol{x}\right) \quad (3.54)$$

with $\mathrm{Tr}[\mathrm{h}] = 0$ and $\nabla^t \mathrm{h} = 0$. As for scalar perturbations we first employ the geodesic equation

$$\frac{d\boldsymbol{p}}{d\eta} = -\frac{p}{2}\nabla(\boldsymbol{l}^t \mathrm{h}\boldsymbol{l})\,, \qquad \frac{d\boldsymbol{x}}{d\eta} \simeq \boldsymbol{l} + \mathcal{O}(\mathrm{h})\,, \quad (3.55)$$

this time for the metric (3.54), and find using

$$\frac{\omega}{T} = \frac{p_0}{T\sqrt{g_{00}}} \simeq \frac{p}{T_0 a}\left(1 - \frac{\delta T}{T} - \frac{1}{2}\boldsymbol{l}^t \mathrm{h}\boldsymbol{l}\right), \quad (3.56)$$

that the Boltzmann equation to first order becomes

$$\left(\frac{\partial}{\partial \eta} + \boldsymbol{l} \cdot \nabla_{\boldsymbol{x}}\right) \frac{\delta T}{T} = -\frac{1}{2} l^t \frac{\partial \mathrm{h}}{\partial \eta} l \ . \tag{3.57}$$

The solution to (3.57), and thus the multipole moments associated with tensor perturbations, which we denote by C_ℓ^T, have in general to be calculated numerically. However, for certain regions of ℓ one finds exact solutions. As for the scalar perturbations, there is a flat plateau at low multipoles due to super-horizon gravitational waves,

$$\ell(\ell+1)\, C_\ell^T = \mathrm{const.} \qquad \text{for } \ell_r \gg \ell \gg 1 \ . \tag{3.58a}$$

At the larger ℓ, roughly where the acoustic peaks occur, one has

$$\ell(\ell+1)\, C_\ell^T \propto \frac{1}{\ell^2} \qquad \text{for } \ell \gg \ell_{\mathrm{eq}} \ , \tag{3.58b}$$

where $\ell_r := \eta_0/\eta_r$ and $\ell_{\mathrm{eq}} := \eta_0/\eta_{\mathrm{eq}}$, which take the values $\ell_r \simeq 55$ and $\ell_{\mathrm{eq}} \simeq 150$ for the concordance model (c.f. [Muk05]). These results mean that, after the plateau at moderately small ℓ, we find a strong decrease at higher ℓ, making thus the tensor contribution to the total C_ℓ power spectrum negligible.

For the concordance model one finds that — even for low multipoles — the tensor modes are always at least one order of magnitude suppressed. Furthermore, from just the temperature multipole moments it is impossible to disentangle the tensor contribution from other effects, like a **spectral tilt**, i.e. $n_s \neq 1$, or reionisation. The study of different polarisation-mode correlations provides a better way to detect primordial gravitational waves.

3.4.4 Polarisation

As concisely discussed in section 3.4.1, the process of recombination is in general not instantaneous, but gradual. Through their interactions with free electrons via Thomson scattering, the CMB photons become *linearly polarised*.

Let us first decompose the electric field $\boldsymbol{E}$ as

$$\boldsymbol{E} = E^a \boldsymbol{e}_a \ , \tag{3.59}$$

where $a \in \{1, 2\}$ labels two orthogonal directions in the plane perpendicular to the direction of propagation. If $\boldsymbol{m} := m^a \boldsymbol{e}_a$ denotes the orientation of polarisation, the **polarisation tensor**, which is defined through

$$\mathrm{P}_{ab} := \frac{\langle E_a E_b \rangle - \frac{1}{2}\langle E_c E^c \rangle\, \boldsymbol{e}_a \cdot \boldsymbol{e}_b}{\langle E_e E^e \rangle} \ , \tag{3.60}$$

3.4 Other Effects

is connected to the brightness-temperature fluctuations via

$$\delta T(\boldsymbol{m}) \propto \mathrm{P}_{ab}\, m^a m^b\,. \qquad (3.61)$$

In this sense, one can determine the polarisation tensor for linear polarisation. Such an electromagnetic wave with electric field $\boldsymbol{E}$ has, after it is scattered in direction $\boldsymbol{n}$ by an electron, the electric field

$$\tilde{\boldsymbol{E}} \propto \bigl((\boldsymbol{E} \times \boldsymbol{n}) \times \boldsymbol{n}\bigr)\,, \qquad (3.62)$$

where we assumed Thomson scattering. For unpolarised light scattered in direction $\boldsymbol{n}$, one has to integrate over all incoming directions $\boldsymbol{l}$ and finds after some algebra (c.f. [Muk05])

$$\mathrm{P}_{ab}(\boldsymbol{n}) \simeq 3 \int \mathrm{d}\eta \int \frac{\mathrm{d}^2 l}{4\pi} \left[\frac{1}{2}\boldsymbol{e}_a \cdot \boldsymbol{e}_b \left(1 - (\boldsymbol{l}\cdot\boldsymbol{n})^2\right) - (\boldsymbol{l}\cdot\boldsymbol{e}_a)(\boldsymbol{l}\cdot\boldsymbol{e}_b)\right] \\ \times \frac{\delta T(\eta,\boldsymbol{l})}{T} \tau'(\eta)\, \mathrm{e}^{-\tau(\eta)}\,, \qquad (3.63)$$

where we used the fact that the intensity is proportional to the fourth power of the temperature. So one infers that the polarisation tensor is quadratic in ℓ and is thus *proportional to the quadrupole temperature anisotropy*, generated during the delayed recombination. Furthermore, it is proportional to the duration of recombination. Numerically it has been shown that polarisation never exceeds 10% of the temperature fluctuations on any scale.

Since P is symmetric and traceless, it has two independent components. It is convenient to decompose it as

$$\mathrm{E}(\boldsymbol{n}) := \mathrm{P}_{ab}{}^{;ab}\,, \qquad (3.64\mathrm{a})$$
$$\mathrm{B}(\boldsymbol{n}) := \mathrm{P}_a{}^{b;ac}\epsilon_{cb}\,, \qquad (3.64\mathrm{b})$$

where the first quantity behaves as a scalar under spatial reflections, being therefore referred to as **electric polarisation mode** (or **E mode**), whereas the second quantity is called the **magnetic polarisation mode** (or **B mode**), due to its odd behaviour under spatial reflections.

Both modes have very distinct polarisation patterns. To illustrate this, let us introduce the **polarisation vector** p_a. It is defined as an eigenvector of the polarisation tensor,

$$\mathrm{P}^a{}_b\, p_a \propto p_b\,. \qquad (3.65)$$

For the E mode one can show that the polarisation vector is proportional to either $\boldsymbol{e}_\theta$ or $\boldsymbol{e}_\varphi$ at every point. The B mode polarisation vectors are linear combinations of $\boldsymbol{e}_\theta$ and $\boldsymbol{e}_\varphi$ and hence orientated in circulating patters around the direction of the $\boldsymbol{k}$.

An important fact is that *scalar perturbations do not generate* B *modes*. Thus, the associated correlation function provides a decisive way to detect primordial gravitational waves. As with the polarisation patterns, the cross correlations

$$C^{ET}(\theta) := \left\langle E(\boldsymbol{n}_1) \frac{\delta T(\boldsymbol{n}_2)}{T} \right\rangle_\theta, \qquad (3.66a)$$

$$C^{BT}(\theta) := \left\langle B(\boldsymbol{n}_1) \frac{\delta T(\boldsymbol{n}_2)}{T} \right\rangle_\theta, \qquad (3.66b)$$

and in turn their associated multipole moments C_ℓ^{ET} and C_ℓ^{BT}, display quite different behaviour. While, for the concordance model, the BT cross correlation decreases above about $\ell \simeq 100$, one finds that the ET cross correlation increases till $\ell \simeq 1000$. From the previous subsection it is clear that around the reionisation horizon, i.e. for $\ell \simeq \ell_{ion}$, one has an extra contribution to polarisation, which is thus able to tremendously amplify the cross polarisation multipole moments around this ℓ.

Hence, the precise knowledge of the polarisation correlation functions allows not only for a distinct study of primordial gravitational waves and of the recombination history, but provides a way to analyse reionisation.

Chapter 4
Stochastic Cosmological Inflation

»Ach, daß die inn're Schöpfungskraft
Durch meinen Sinn erschölle!
Daß eine Bildung voller Saft
Aus meinen Fingern quölle!

Wirst alle meine Kräfte mir
In meinem Sinn erheitern
Und dieses enge Dasein hier
Zur Ewigkeit erweitern.«

JOHANN WOLFGANG VON GOETHE

Inflationary cosmology has become a successful building block of our current understanding of the Universe. While descriptions at the classical and semi-classical level seem fairly well developed, a full understanding of the inflationary dynamics, including regions of strong quantum fluctuations (c.f. figure 4.1), is still lacking. A major step in this direction has been made by Starobinsky [Sta86], who introduced the concept of stochastic inflation, which provides a framework to study the evolution of quantum fields in an inflationary Universe [GLM87, Rey87, Kan89, LLM94]. This approach has acquired considerable interest over the last few years [WV00, Bel00, Bel01, LMM+04, HY05, MM05, TW05, MB06, MM06b, MM06a, LW07, KS08, KS09]. Its idea lies in splitting the quantum fields into long- and short-wavelength modes, and viewing the former as classical objects evolving stochastically in an environment provided by quantum fluctuations of shorter wavelengths.

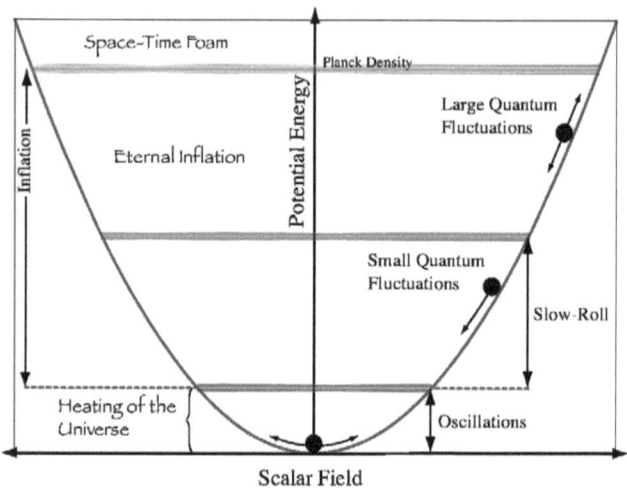

Figure 4.1: Schematic picture of the stages of inflation via a scalar field.

Given the de Sitter horizon as a natural length scale of the problem, one then focusses on the 'relevant' degrees of freedom (the long-wavelength modes) and regards the short-wavelength modes as 'irrelevant' ones, where 'short' and 'long' are due to the horizon.

This chapter is organised as follows: First we derive an effective stochastic equation of motion for the long-wavelength field, which in turn is solved in detail for simple cases and approximations. Then, we give an overview of major results obtained in the literature, including a sketch of a rigorous approach to stochastic inflation.

4.1 Effective Equation of Motion

Let us in part summarise and follow Rey ([Rey87]) and take a minimally-coupled real scalar field with the Lagrangian

$$\mathcal{L} = \frac{1}{2} g^{\mu\nu} \partial_\mu \Phi \partial_\nu \Phi - V(\Phi) , \qquad (4.1)$$

where Φ is a gauge singlet and V its potential. We assume a fixed de Sitter background geometry, i.e. we take $(g_{\mu\nu}) = \mathrm{diag}(1, -a(t)^2, -a(t)^2, -a(t)^2)$ with the scale factor $a(t) = \exp(t)$. As before we set $\hbar \stackrel{!}{=} c \stackrel{!}{=} 1$ but $8\pi G \neq 1$ and instead use $H \stackrel{!}{=} 1$ for convenience.

4.1 Effective Equation of Motion

Then, with the definition

$$\Box := \frac{1}{\sqrt{g}} \partial_\mu \left(\sqrt{g}\, g^{\mu\nu} \partial_\nu \right), \qquad (4.2)$$

where g is the absolute value of the determinant of the metric tensor ($g_{\mu\nu}$), one finds the exact equation of motion

$$\Box \Phi + \frac{\partial V(\Phi)}{\partial \Phi} = 0, \qquad (4.3a)$$

or

$$\left(\frac{\partial^2}{\partial t^2} + 3 \frac{\partial}{\partial t} - e^{-2t} \nabla^2 \right) \Phi + \frac{\partial V(\Phi)}{\partial \Phi} = 0. \qquad (4.3b)$$

To derive an effective equation of motion, one performs a **split** of the full quantum field Φ **into a short- and a long-wavelength** part like

$$\Phi = \phi + \varphi, \qquad (4.4)$$

where ϕ and φ contain (in the Fourier sense) the wavelengths that are shorter and longer than the Hubble scale, respectively. Specifically, the splitting is done by a filter function W, which might be approximated by a sharp cut, i.e. $W = \Theta$. In general, one has

$$\phi(t, \boldsymbol{x}) \simeq \int d^3k\, W\!\left(k\, \eta(t) - \epsilon\right) \left[\hat{a}(\boldsymbol{k})\, u(t, k)\, e^{-i\boldsymbol{k}\cdot\boldsymbol{x}} + \hat{a}^\dagger(\boldsymbol{k})\, u^*(t, k)\, e^{+i\boldsymbol{k}\cdot\boldsymbol{x}} \right], \qquad (4.5)$$

where the corrections of this free-field expansion to the full case vanish at leading order in the Taylor coefficients of V, which are anyway constraint to be small. The parameter ϵ basically says where to split into short and long wavelengths, and should be chosen much smaller than one in order to cut far beyond the Hubble radius. On the other hand, this parameter should not be to small, because otherwise the change in the background metric has to be taken into account. One can show [Sta82] that the bounds on this parameter can be expressed via $1 \gg \epsilon \gg \exp\!\left(- H^2/\dot{H}\right)$.

The mode function obeys the equation of motion for free fields, i.e. it is a solution of equation (4.3a) with $V = 0$, and is approximately given by

$$u(t, k) \simeq \frac{1}{\sqrt{2k}} \left(\eta(t) - i \frac{\eta(t)}{k} \right) e^{-i k \eta(t)}. \qquad (4.6)$$

We define the **noise** ζ via

$$\zeta(t, \boldsymbol{x}) := \frac{1}{3} \left(\frac{\partial^2}{\partial t^2} - 3 \frac{\partial}{\partial t} + e^{-2t} \nabla^2 \right) \phi(t, \boldsymbol{x}), \qquad (4.7)$$

and further

$$v[\varphi] := \frac{1}{3}\int d^3x\, dt \left[\frac{1}{2}\left(e^{-t}\nabla\varphi\right)^2 + V(\varphi)\right]. \tag{4.8}$$

Assuming slow-roll, we obtain a **functional Langevin equation** for the coarse-grained field φ:

$$\frac{\partial\varphi(t,\boldsymbol{x})}{\partial t} \simeq -\frac{\delta v[\varphi]}{\delta\varphi(t,\boldsymbol{x})} + \zeta(t,\boldsymbol{x}) . \tag{4.9}$$

As it stands, it is viewed as a *stochastic* equation of motion for the *classical* field φ. For a full specification one needs to specify the noise correlators of ζ, which are basically determined by the quantum averages of the short-wavelength field ϕ.

For a sharp cut, a simple calculation shows

$$\zeta(t,\boldsymbol{x}) \simeq i\frac{\epsilon}{\eta(t)} \int d^3k\, \delta\big(\eta(t)k - \epsilon\big) \sqrt{\frac{1}{2k^3}} \left[\hat{a}(\boldsymbol{k})\,e^{-i\boldsymbol{k}\cdot\boldsymbol{x}} + \hat{a}^\dagger(\boldsymbol{k})\,e^{+i\boldsymbol{k}\cdot\boldsymbol{x}}\right], \tag{4.10}$$

from which one can check that

$$[\zeta(t,\boldsymbol{x}),\zeta(t,\boldsymbol{y})] \simeq 0 , \tag{4.11}$$

hence justifying the 'classical' treatment of 'quantum' operators. It follows that ζ is a Gaussian white noise with two-point correlator (c.f. [Rey87])

$$\begin{aligned}\Delta(t,\boldsymbol{x},t',\boldsymbol{y}) &:= \langle\zeta(t,\boldsymbol{x})\zeta(t',\boldsymbol{y})\rangle \\ &\simeq \frac{1}{4\pi}\frac{\sin(\epsilon|\boldsymbol{x}-\boldsymbol{y}|/\eta(t))}{\epsilon|\boldsymbol{x}-\boldsymbol{y}|/\eta(t)}\delta(t-t') ,\end{aligned} \tag{4.12}$$

which also displays the Markovian nature of this random process. Of course, this property heavily relies on the choice of the filter function — an issue which is studied in detail in section 6.3.

It is easy to show (c.f. [Rey87]) that the stochastic Langevin equation (4.9) is equivalent to a **Fokker-Planck equation** for the probability density functional $\mathcal{P}[\varphi,t]$, namely

$$\begin{aligned}\frac{\partial\mathcal{P}[\varphi,t]}{\partial t} &\simeq \int d^3x\, \frac{\delta}{\delta\varphi(t,\boldsymbol{x})}\left[\frac{\delta v[\varphi]}{\delta\varphi(t,\boldsymbol{x})}\mathcal{P}[\varphi,t]\right] \\ &+ \frac{1}{2}\int d^3x\, d^3y\, \Delta(t,\boldsymbol{x},t',\boldsymbol{y})\frac{\delta}{\delta\varphi(t,\boldsymbol{x})}\frac{\delta}{\delta\varphi(t,\boldsymbol{y})}\mathcal{P}[\varphi,t] .\end{aligned} \tag{4.13}$$

The distribution $\mathcal{P}[\varphi,t]$ is defined such that the statistical average of a functional $\mathcal{F}[\varphi]$ can be expressed as

$$\langle\mathcal{F}[\varphi]\rangle \equiv \int \mathcal{D}[\varphi]\, \mathcal{F}[\varphi]\mathcal{P}[\varphi,t] . \tag{4.14}$$

4.1 Effective Equation of Motion

It also obeys the normalisation condition

$$\int \mathcal{D}[\varphi]\, \mathcal{P}[\varphi, t] \overset{!}{=} 1 \,. \tag{4.15}$$

Solutions to the Fokker-Planck equation (4.13) for an inhomogeneous field can become rather complicated. In fact, this is an unsolved problem in general, and hence we will now illustrate the power of the stochastic formalism for homogeneous fields. An extension to spatial dependencies is developed in the next two chapters.

Homogeneous Solutions

Following [Rey87] let us now derive an analog to (4.13) for homogeneous fields and discuss corresponding solutions. It is interesting to note that below the coarse-graining scale $\sim H^{-1}$, one finds that the quantity

$$F(\varphi, t) := \exp\left(\frac{4\pi^2}{3}(V(\varphi) - V(0))\right) \mathcal{P}(\varphi, t) \tag{4.16}$$

obeys the Schrödinger-like equation

$$-\frac{\partial F(\varphi, t)}{\partial t} \simeq \hat{H}_{\text{FP}} F(\varphi, t) \,, \tag{4.17}$$

with the **Fokker-Planck Hamiltonian**

$$\hat{H}_{\text{FP}} := -\frac{1}{2}\left[\frac{1}{4\pi^2}\frac{\partial^2}{\partial \varphi^2} + (V'(\varphi))^2 - V''(\varphi)\right] \,. \tag{4.18}$$

To calculate the moments $\langle \varphi^n \rangle$ we employ (4.13) and use $\mathcal{P}(\varphi \to \pm\infty, t) = 0$ to obtain (c.f. [Rey87])

$$\begin{aligned}
\frac{\partial}{\partial t}\langle \varphi^n \rangle &= \int_{-\infty}^{+\infty} d\varphi\, \varphi^n \frac{\partial \mathcal{P}(\varphi, t)}{\partial t} \\
&\simeq \frac{1}{8\pi^2}\int_{-\infty}^{+\infty} d\varphi\, \varphi^n \frac{\partial^2 \mathcal{P}(\varphi, t)}{\partial^2 \varphi} + \frac{1}{3}\int_{-\infty}^{+\infty} d\varphi\, \varphi^n \frac{\partial}{\partial \varphi}\left[\frac{\partial V}{\partial \varphi} \mathcal{P}(\varphi, t)\right] \\
&= \frac{1}{8\pi^2} n(n-1)\langle \varphi^{n-2} \rangle - \frac{n}{3}\left\langle \varphi^{n-1} \frac{\partial V}{\partial \varphi}\right\rangle \,.
\end{aligned} \tag{4.19}$$

For a **free massive field** with the potential

$$V(\varphi) = \frac{1}{2}\mu^2 \varphi^2 \,, \tag{4.20}$$

one has (c.f. [Rey87])

$$\left(\frac{\partial}{\partial t} + \frac{2\mu^2}{3}\right)\left\langle\varphi^2(t)\right\rangle \simeq \frac{1}{4\pi^2} \tag{4.21}$$

and finds the result (c.f. [Rey87])

$$\left\langle\varphi^2(t)\right\rangle \simeq \frac{3}{8\pi^2\mu^2}\left[1 - \exp\left(-\frac{2\mu^2 t}{3}\right)\right] \xrightarrow{\mu\to 0} \frac{t}{4\pi^2}. \tag{4.22}$$

A **Higgs-type potential**

$$V(\varphi) = \frac{\lambda}{4}\left(\varphi^2 - \frac{\mu^2}{\lambda}\right)^2 \tag{4.23}$$

with $\lambda > 0$ leads to (c.f. [Rey87])

$$\left(\frac{\partial}{\partial t} - \frac{2\mu^2}{3}\right)\left\langle\varphi^2(t)\right\rangle \simeq \frac{1}{4\pi^2} - \frac{2\lambda}{3}\left\langle\varphi^4(t)\right\rangle. \tag{4.24}$$

In general, this equation is difficult to solve exactly, but with the **mean-field** (Hartree-Fock) **approximation**,

$$\left\langle\varphi^4(t)\right\rangle \simeq 3\left\langle\varphi^2(t)\right\rangle^2, \tag{4.25}$$

one easily obtains (c.f. [Rey87])

$$\left\langle\varphi^2(t)\right\rangle \simeq \frac{3}{8\pi^2\mu^2}\left[\exp\left(\frac{2\mu^2 t}{3}\right) - 1\right] + f(t), \tag{4.26a}$$

where

$$f(t) \simeq \begin{cases} \frac{3\lambda}{\mu^2}\left(\frac{3}{8\pi^2\mu^2}\right)^2\left[\frac{2\mu^2 t}{3} - \left(\frac{2\mu^2 t}{3}\right)^2\right] & : \ t \ll 1, \\ -\frac{2\lambda t^2}{8\pi^2}\exp\left(\frac{2\mu^2 t}{3}\right) & : \ t \gtrsim 1. \end{cases} \tag{4.26b}$$

This displays the exponential acceleratory behaviour of the scalar field as it rolls down the potential hill. One easily checks that in the limit $\lambda \to 0$ and for $\mu^2 \to -\mu^2$ the solution (4.26a) reproduces (4.22).

In fact, for many scalar fields there is another approximation, namely that of large N. Therefore, one takes an N-**component scalar field** $\vec{\varphi}$ evolving in the potential

$$V(\vec{\varphi}^2) = \frac{\lambda'}{4}\left(\vec{\varphi}^2 - \frac{\mu^2}{\lambda'}\right)^2, \tag{4.27}$$

with $\lambda' := \lambda/N$ and $\lambda \in \mathbb{R}$ fixed. Then a simple analysis [Rey87] shows that

$$\left\langle \vec{\varphi}^2(t) \right\rangle \simeq \frac{3N}{8\pi^2\mu^2}\left[1 - \exp\left(-\frac{2\mu^2 t}{3}\right)\right] + \mathcal{O}\left(\frac{1}{N}\right). \tag{4.28}$$

Except for the amplification by the factor N and up to $\mathcal{O}(N^{-1})$-corrections, it is essentially the result of a simple free massive scalar field derived in (4.22).

4.2 Other Results

After rather detailed derivations in the previous section, we will now give a brief over-view over some major results that have been obtained so far.

Let us proceed chronologically. As we have already mentioned the fundamental 1986-proceedings article by Starobinsky and discussed in detail the paper by Rey, we would like to shortly comment on the 1988-paper by Nakao et al. [NNS88]. There, the authors studied stochastic dynamics in the context of the scenario of **new inflation**, where all the energies involved are well below the Planck mass. This means, they took for the early stage of inflation a Higgs-type double-well potential of the form (4.23), where they assumed

$$\frac{\mu^4}{m_{\text{Planck}}^4} \ll \lambda \ll \frac{\mu^2}{m_{\text{Planck}}^2}, \tag{4.29}$$

with m_{Planck} being the Planck mass. They put special emphasis on the condition for the realisation of the slow roll-over phase, in which the scalar field evolves according to the slow-roll version of the classical equation of motion. When determining this from the comparison of the quantum force to the potential force, they took into account the difference in the physical volume of a horizon-size region, because of the φ-dependence of the expansion rate H. A result is that on a global scale, the Universe would expand forever till the scalar field reaches a critical value φ_* in some region and begins to evolve into one of the potential minima. Then this region enters the classical slow roll-over phase. The authors determined this critical value to be $\varphi_* \simeq 0.5 \pm 0.1\, H^3 \mu^{-2}$.

This set-up of **eternal inflation**, in which the Universe — having neither beginning nor end — constitutes of regions that expand forever, has already been studied in 1983 by Vilenkin [Vil83] also using a stochastic formalism.

In [MOL89] Matarrese et al. studied the stochastic dynamics of the inflaton for a variety of models (free massive, quartic massless, exponential). They found that for scale-free potentials a so-called **scaling regime** arises, where the coarse-grained probability distribution develops a sharp peak around the classical slow-roll solution. This solution is non-Gaussian and a self-similar function of a single scaling variable. It is also worth noting that the authors have a discussion on Itô and Stratonovich calculus and include both as special cases in their derivations.[1]

Then in 1994, a major step towards a deeper understanding and a better calculability was made with the paper [SY94] by Starobinsky and Yokoyama. The authors studied in detail the stochastic evolution of a slowly-rolling scalar test field in exact de Sitter space-time. They developed a general method of calculating arbitrary n-point correlation functions, which allowed them to prove the de Sitter invariance for a self-interacting scalar field with a small mass term of arbitrary sign. It has been found that these fields loose their memory of their initial state at the beginning of the de Sitter phase at late times. For the Higgs-type potential (4.23) the corresponding relaxation time has been calculated to

$$t_{\text{rel}} \sim \min\left\{ H \left|\mu\right|^{-2}, H^{-1} \lambda^{-1/2} \right\}, \tag{4.30}$$

which can be much shorter than the correlation time. For the double-well potential (4.23) with $|\mu|^2 \gg \sqrt{\lambda}\, H^2$, the space is covered by domains with $\varphi = \pm\varphi_0$, where φ_0 is the absolute value of a minimum of (4.23). These domains are separated by relatively thinner (although being still much thicker than H^{-1}) domain walls between them. Their typical size is

$$R_{\text{DW}} \sim H^{-1} \exp\left(H^2 / |\mu|^2 \right), \tag{4.31a}$$

that of these domains is found to be

$$R_{\text{D}} \sim H^{-1} \exp\left(\pi^2 \mu^4 / 3 \lambda H^4 \right). \tag{4.31b}$$

First we notice that $R_{\text{DW}} \ll R_{\text{D}}$, i.e. the domain walls are thin walls under the given assumptions, and second that the *physical* lengths R_{DW} and R_{D} are independent of time.

This means that their *comoving* size is shrinking. Hence, the expansion of the background space-time is not followed by a corresponding expansion of physical domains. Instead new domains, separated by new domain walls, are created at a constant rate.

A careful study of the spatial dependence of the noise correlator has been performed by Winitzki and Vilenkin [WV00]. In their paper they put special emphasis on possible smooth filter functions.

The investigation of the wave function of the matter field fluctuations in the infra-red sector is dealt with in the paper [Bel01] by Bellini. For power-law inflation, where $a(t) \propto t^p$, it has been found

---

[1] At this point we would like to recommend the excellent article *Itô versus Stratonovich* [van81].

4.2 Other Results

from a phase-space analysis with focus on coherence aspects, that for $p < 4.6$ a classical stochastic treatment is not valid.

In the same spirit — concerning the reliability of stochastic inflation — Martin and Musso [MM06b] presented a new method to estimate the precision of the perturbative expansion studied in their previous paper [MM06a]. It has been shown with this method, which is based on the use of the Lagrange remainder theorem [WW90], that except for the very end of inflation, the approximate probability density functions derived in [MM06a] are very good approximations to the actual ones.

More on the application side of the stochastic formalism is the paper [LMM+04] by Liguori et al. In this work, the authors generalise the treatment of inflationary perturbations to consider non-Markovian coloured noise, and calculate the power spectrum of the gauge-invariant comoving curvature perturbation to first oder in the slow-roll parameters. For this quantity they found a blue tilt on the largest observable scales.

In 2005 a rigorous derivation and an extension (to scalar QED) of the stochastic inflation framework has been given by Woodard [Woo05]. Let us briefly sketch his line of thought. We note that in general, non-quadratic interactions result in non-linear noise terms on the right hand side of the field equation. However, if one is interested in the late-time behaviour, or more precisely in the leading-$\ln(a(t))$ contribution, one may restrict to linear, Gaussian-distributed noise terms. As discussed, this has been argued already a long time ago by Starobinsky [Sta82] and has been rigorously proven by Woodard [Woo05] (see also [TW05, MW06]). Therefore, consider the model (4.1) with the potential $V(\varphi) = \frac{1}{4}\lambda\varphi^4$ in de Sitter background. Then the energy density ε reads

$$\varepsilon(t) = \gamma + \frac{3\lambda}{32\pi^4}\left[-\frac{1}{2}\ln^2\big(a(t)\big) + \zeta \ln\big(a(t)\big) + \mathcal{O}\big(a(t)^{-1}\big)\right] + \mathcal{O}\big(\lambda^2\big), \qquad (4.32)$$

with $\gamma, \zeta \in \mathbb{R}$. The quantity γ contains all possible constant terms, in particular the one-loop effect from the kinetic energy of the inflationary scalar.

In general, when calculating observables in scalar-field theories in de Sitter background, one generically encounters so-called **infra-red logarithms**, which means factors of $\ln(a(t)) = t$. These grow without limit until they overcome the (self-)coupling and eventually become non-perturbative. Hence, new methods are needed to tackle those issues. One is the above-mentioned **leading-log approximation** [Woo05].

From the conjecture stating that powers of $a(t)^{-1} = e^{-t}$ can be resumed by a renormalisation of the initial state [OW02], it follows the energy density can be expressed in the general form

$$\varepsilon(t) = \text{const.} + \sum_{n=2}^{\infty}\lambda^{n-1}\sum_{k=0}^{2n-3}c_k^{(n)}\Big[\ln\big(a(t)\big)\Big]^{2n-2-k}, \qquad (4.33a)$$

where the $c_k^{(n)}$s are real numbers. The leading-log approximation would be in this case

$$\varepsilon(t)\Big|_{\text{leading-log}} = \text{const.} + \sum_{n=2}^{\infty} c_0^{(n)} \left[\lambda \ln\big(a(t)\big)^2\right]^{n-1}, \qquad (4.33b)$$

which takes into account arbitrary powers of λ but only the dominant (late-time) contribution in the scale factor. If one works with the infra-red field φ, i.e. the part of Φ which is separated from the high-frequency modes by the filter function W, one sees that the field equation changes. This in turn does not leave the original stress-energy tensor conserved because of ultra-violet modes that redshift past the horizon. As shown in [Woo05], it is indeed possible — by adding non-local source terms — to modify $T^{\mu\nu}$ to obtain a fully consistent model of only the infra-red modes which reproduces the leading infra-red logarithms. Employing then the Yang-Feldman equation [YF50] makes it possible to derive the stochastic Langevin equation (4.9).

Chapter 5
Replica Field Theory

»*Aus der unbekannten Ferne*
von des ew'gen weiten Zelt
leuchtet hier ein kleiner Sterne –
ein Gedanke in die Welt.«

according to JULIUS LOHMEYER

So far we discussed some basic concepts of cosmology with the focus on cosmological inflation. In the last chapter we summarised the ideas behind stochastic inflation which have been developed so far. As the aim of this thesis is to go beyond some of the mentioned aspects, we would like do devote this chapter to the fundamentals of replica field theory which actually allows us to extend the progress made in the past. This presentation shall be done in a concise stand-alone manner, with an introduction and an explanation of these methods in the context they were originally developed for, namely so-called disordered systems. However, the way of presenting is chosen to be directly applicable to stochastic inflation. This application will be performed in the following chapter.

The treatment of disordered systems is an important task since almost all real systems are not really pure, i.e. free from impurities. Examples of such systems in condensed-matter physics that have been intensively studied are amorphous magnets [HPZ73, SO92], liquid crystals in porous media [CBM+93], nematic elastomers, He-3 in aerogel [PP95] and vortex phases of impure superconductors [BFG+94]. The presence of random impurities produces in general an energy landscape with plenty of meta-stable states which makes the determination of the global ground state highly non-trival. Such states are very close in energy but in phase-space they may be far apart.

The structure of this chapter is as follows. First we discuss the perhaps most famous example of a disordered system: The random-field Ising model. Second, we introduce pure elastic systems and discuss the effect of disorder. This is followed by the introduction of a method to perform disorder averages: The replica trick. The chapter continues with a section on disorder distributions. We conclude with a discussion of the Gaussian variational method which is followed by two sections on replica symmetry breaking, including mathematical details and a worked example.

5.1 Random Field Ising Model

Probably the oldest, simplest, discrete non-trivial model which exhibits spontaneous symmetry breaking is the **Ising model** (c.f. [Nat97, Bel98] for reviews). In the presence of an external field H, its Hamiltonian is given by

$$\mathcal{H} = -\sum_{<ij>} J_{ij} u_i u_j + \sum_i H_i u_i , \qquad (5.1)$$

with $u_i = \pm 1$ for all lattice sites i, and $\sum_{<ij>}$ means summation over nearest neighbours. Impurities in the quantities J_{ij} and H_i are modelled by additional random contributions:

$$J_{ij} \rightarrow J_{ij} + \delta J_{ij} , \qquad H_i \rightarrow H_i + h_i . \qquad (5.2)$$

In a simple case, h_i and δJ_{ij} are Gaussian-distributed random variables with mean zero:

$$\overline{\delta J_{ij} \delta J_{kl}} = S_{ijkl} J^2 , \qquad \overline{h_i h_j} = \delta_{ij} h^2 , \qquad (5.3)$$

with

$$S_{ijkl} := \frac{1}{3} \left(\delta_{ij} \delta_{kl} + \delta_{ik} \delta_{jl} + \delta_{il} \delta_{jk} \right) . \qquad (5.4)$$

When $\delta J = 0$ and $\delta h \neq 0$ we talk about pure **random-bond** disorder (or noise). The situation with $\delta J \neq 0$ and $\delta h = 0$ is referred to as the pure **random-field** case. Assuming

$$\delta J_{ij} = 0 , \qquad h_i \neq 0 , \qquad h \ll J , \qquad (5.5)$$

and taking the continuum limit, one can show that the above theory can be modelled by the so-called **Ginzburg-Landau functional**

$$\mathcal{H}[s] = \frac{1}{2} \int_x \left(\nabla_x u(x) \right)^2 - \frac{\mu^2}{2} \int_x \left(u(x) \right)^2 + \frac{\lambda}{4} \int_x \left(u(x) \right)^4 + \int_x \mathrm{h}(x) s(x) , \qquad (5.6)$$

5.1 Random Field Ising Model

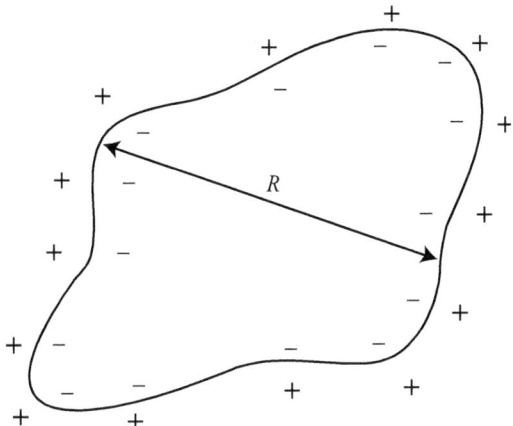

Figure 5.1: A domain of reversed spins of size R.

where $\int_x := \int d^d x$, $\lambda > 0$, and $\mu^2, \lambda \in \mathbb{R}$. Equation (5.3) then becomes

$$\overline{h(x)\,h(y)} = \delta(x-y)\,h^2 \;. \tag{5.7}$$

As was shown by an exact treatment (c.f. [Fro84, Ber85]), for sufficiently large random fields h (compared to J), the system is disordered at low temperatures. For the case of h $\ll$ J and $d \leq 2$ it has been proven [IM75] that the ground state becomes unstable with regard to the formation of ill-oriented domains (see figure 5.1).

There are also some experiments on the random-field Ising Model. Based on univer-sality, there is no need that the Hamiltonian of the studied system does *precisely* correspond to that of this model — the important aspect is that it has the appropriate *symmetries*. Perhaps the most intensively studied experimental realisations are diluted antiferromagnets in a homogeneous external field [FA79, Bel98]. There, the combination of the external field and the dilution yields a random field. Other examples are adsorbed mono-layers on impure substrates like Cu(100) or Xe [Vil82], mixed Jahn-Teller liquids [deG84, DW87] or binary liquids in porous media [GMP+87]. Gutin et al. discussed cooperativity of protein folding as an application of the random-field Ising model [GAS96].

5.2 Elastic Systems

Elastic systems play a crucial role in statistical physics. In a discrete model where a displacement at lattice site i is described by the variables $u_i \in \mathbb{R}$, the Hamiltonian might be approximated by

$$\mathcal{H} = \frac{k}{2} \sum_{j,z} (u_j - u_{j+z})^2 , \qquad (5.8)$$

where $k \in \mathbb{R}$ is the **elasticity constant** and z spans over nearest neighbours. Assuming that the quantity u_j varies smoothly at low-enough temperatures, we can go to a continuous variable $u(x)$ with

$$u_{j+z} - u_j \to z \cdot \nabla u(x) , \qquad (5.9)$$

and find thus for a d-dimensional lattice

$$\mathcal{H} = \frac{k}{2} \int \mathrm{d}^d x \sum_{a,b} \sum_z z^a z^b \, \partial_a u(x) \, \partial_b u(x) . \qquad (5.10)$$

On a square lattice with lattice constant α the result is [Gia03]

$$\mathcal{H} = \frac{k\alpha^2}{2} \int \mathrm{d}^d x \sum_a \partial_a u(x) \, \partial_a u(x) . \qquad (5.11)$$

Adding Disorder

The above considerations were made for pure systems. We now discuss the inclusion of disorder. Perhaps the simplest physical example of a disordered system is an Ising magnet with impurities at low temperature T. Imposing boundary conditions such that all spins are up on the upper and all spins are down on the lower boundary, a domain wall separates the two regimes. It is completely flat at $T = 0$. Disorder, e.g. due to wrong or missing constituents, can deform this wall — it becomes *rough* (c.f. figure 5.2).

As in the previous section, the displacement is described by a scalar function. An example with a more-component displacement is the Bragg glass. Charge-density waves are three-dimensional examples. So each of the systems can be described by an N-component displacement-vector field $\vec{\varphi}(x)$. Usually, the number N is called **external dimension** and the number d of components of x is labelled as the **internal dimension** of the underlying manifold.

5.3 Replica Trick

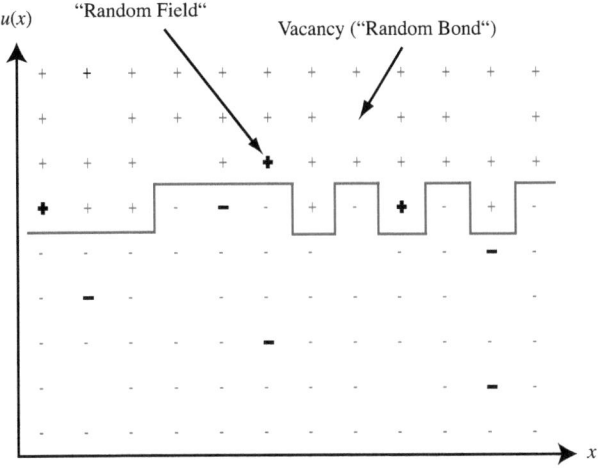

Figure 5.2: Deformation of a domain wall of an Ising magnet by disorder at low temperature. Without impurities it is flat.

We already saw that *pure* elastic systems can be described by

$$\mathcal{H}_0 = \frac{1}{2} \int_x \sum_a \partial_a \vec{\varphi}(x) \cdot \partial_a \vec{\varphi}(x) \,, \tag{5.12}$$

where we absorbed the constants in the definition of the fields. Since we now deal with *disordered* systems, we have an additional disorder-induced contribution to the energy, which might be described by a potential V via

$$\mathcal{H}_D = \int_x V(\vec{\varphi}(x)) = \int_x \sum_{j=1}^{\infty} \sum_{i_1...i_j}^N h_{i_1...i_j}(x) \, \varphi_{i_1}(x) \cdots \varphi_{i_j}(x) \,, \tag{5.13}$$

here the (stochastic) Taylor coefficients h_λ depend on the particular disorder realisation.

5.3 Replica Trick

As every observable is an averaged quantity, we have for disordered systems an extra average over the different disorder realisations, which we indicate by a bar. One has in particular to compute the disorder average of the free energy $\overline{\mathcal{F}} = -T \overline{\ln(\mathcal{Z})}$. But this is difficult to achieve directly. One uses instead the so-called **replica trick**

$$\overline{\mathcal{F}} = -T\,\overline{\ln(\mathcal{Z})} = -T\lim_{n\to 0}\frac{1}{n}\ln\left(1 + n\,\overline{\ln(\mathcal{Z})}\right)$$
$$= -T\lim_{n\to 0}\frac{1}{n}\ln\overline{\left(\exp(n\,\ln(\mathcal{Z}))\right)} = -T\lim_{n\to 0}\frac{1}{n}\ln\left(\overline{\mathcal{Z}^n}\right). \quad (5.14)$$

This means, one first has to compute $\overline{\mathcal{Z}^n}$ for n integer. Then, if the result is an analytic function of n, one performs an analytic continuation and takes the limit $n \to 0$. Explicitly:

$$\overline{\mathcal{Z}^n} = \int \prod_{a=1}^{n} \mathcal{D}[\vec{\varphi}_a]\, \overline{\exp\left(-\beta \sum_{b=1}^{n} \mathcal{H}[\vec{\varphi}_b]\right)}$$
$$\equiv \int \prod_{a=1}^{n} \mathcal{D}[\vec{\varphi}_a]\, \exp\left(-\beta \mathcal{H}^{(n)}[\vec{\varphi}_1, \ldots, \vec{\varphi}_n]\right). \quad (5.15)$$

The n-times replicated systems get coupled by the disorder average. In particular a Gaussian disorder distribution yields for the **replicated Hamiltonian**

$$\mathcal{H}^{(n)} = \frac{1}{2}\sum_{a=1}^{n}\int_k G_0^{-1}(k)\,\vec{\varphi}_a(k)\cdot\vec{\varphi}_a(-k) - \int_x f\bigl(\vec{\varphi}_1(x),\ldots,\vec{\varphi}_n(x)\bigr), \quad (5.16)$$

where G_0 is the free propagator and we defined $\int_k := \int d^d k/(2\pi)^d$. The function f is determined by the disorder distribution (see below).

The replica trick is much older than one might think. Marc Mézard said: »*Giorgio Parisi dates it back to at least the fourteenth century when the bishop of Lisieux Nicolas d'Oresme used a similar trick in order to define non-integer powers.*« [Mez04] The replica method allows one to study the free-energy landscape and, in particular, the regions of low free energy. After averaging over disorder, the replicated action, which has been an action of n non-interacting theories, becomes an action without disorder, but with an attractive interaction between the replicas. But why is it attractive? Well, the reason is that they share the same Hamiltonian. Hence they will be attracted towards the same favourable regions in phase space, while they are repelled from unfavourable ones.

In a very simple case, in which the phase space has basically one large valley, the replicas fall altogether in this valley, in other words it is **replica symmetric**. Then the order parameter, the typical 'distance' between any two replicas, gives the valley size. The situation is much more complicated if the system has many meta-stable states. In general, different replicas do not fall into the same valley and the situation one is confronted with is called **replica symmetry breaking**, which is technically spontaneous symmetry breaking where the underlying group is the permutation group. This may look simpler than it is, because one has to take the 'physical' limit $n \to 0$ at the end, which can be rather

5.4 Noise Distributions

non-trival. It turns out that the order parameter is in general a function $P(q)$, which is the probability of finding two replicas having their **overlap** equal to q. Here, the overlap q measures the 'distance' between two replicas. A mathematical specification is presented in the appendix.

5.4 Noise Distributions

In this sections we present some general considerations on the distribution of a random potential. We will formulate them in a way in which they are applicable to stochastic inflation, where the quantum fluctuations induce the noise, c.f. the equation (4.9). In turn, the coordinates are now elements of a space-time manifold. For simplicity we assume zero mean and for the second cumulant of the random potential V the form

$$\overline{V(\vec{\varphi}_a(x),x)\, V(\vec{\varphi}_b(y),y)} = \phi(x,y) N \operatorname{R}\left(\frac{\Xi_{ab}(x,y)}{N}\right), \tag{5.17}$$

where, for later convenience, we rescale by the number N of field components. The function R reflects the correlation in replica field space, with, e.g.

$$\Xi_{ab}(x,y) := \vec{\varphi}_a(x) \cdot \vec{\varphi}_b(y) \tag{5.18a}$$

or

$$\Xi_{ab}(x,y) := \left[\vec{\varphi}_a(x) - \vec{\varphi}_b(y)\right]^2. \tag{5.18b}$$

We refer to (5.18a) as **product correlation**, while (5.18b) is called **difference correlation**. Below we will mainly use the former. Of course, the specific form of the function R, as well as of its argument, have to be determined from first principles. The space-time correlation $\phi(x,y)$ is called **short-range**, if $\phi(x,y) = \wp(t,t') \delta^{d-1}(\boldsymbol{x}-\boldsymbol{y})$, and **long-range** for all other cases. Note, that for the example of a free field, the Gaussian distribution is exact and that the function R is linear. A difference between (5.18a) and (5.18b) is that the latter has the so-called **statistical tilt symmetry**

$$\vec{\varphi}_a(x) \rightarrow \vec{\varphi}_a(x) + \vec{g}(x), \tag{5.19}$$

where $\vec{g}(x)$ is some function without replica index.

When performing the noise average of correlation functions, one has to average the factor $\exp\big(-\sum_a \mathcal{S}[\vec{\varphi}_a]\big)$, where here and in remainder of this chapter, we study the Wick-rotated system. Assuming short-range product correlation, the part containing noise may be calculated as

$$\overline{\exp\left(\sum_\lambda h_\lambda \varphi_\lambda \ldots \varphi_\lambda\right)} \propto \int \mathcal{D}[\{h\}] \exp\left(\sum_\lambda h_\lambda \varphi_\lambda \ldots \varphi_\lambda - \frac{1}{2} \sum_{\lambda,\lambda'} h_\lambda A_{\lambda,\lambda'} h_{\lambda'}\right)$$

$$\propto \exp\left(\frac{1}{2}\sum_{\lambda,\lambda'}\varphi_\lambda \ldots \varphi_\lambda A^{-1}{}_{\lambda,\lambda'}\varphi_\lambda \ldots \varphi_{\lambda'}\right) \tag{5.20}$$

$$\equiv \exp\left(\frac{1}{2}\sum_{a,b=1}^{n}\int_x N\mathrm{R}\left(\frac{\vec{\varphi}_a(x)\cdot\vec{\varphi}_b(x)}{N}\right)\right),$$

where the matrix $(A_{\lambda,\lambda'})$ describes the Gaussian noise distribution.

This is perhaps a good place to mention a difference between what is described in this thesis and what is described in the field of disordered systems. In the latter one studies macroscopic objects, like a crystal with defects, such as vacancies, or wrong constituents, misaligned layers or substrate impurities. These kinds of disorder are mimicked by random variables.

Averages over disorder are thought of as averages over different realisation, i.e. practically different pieces of a crystal, for instance. Mézard and Parisi [MP91] made a consistent replica field theoretic approach to those systems, whose n copies — arising due to the application of the replica trick — are viewed as respectively *different*, coming from different realisations, carrying different random variables, with distinct correlations among them.

This is, however, *not* the case in stochastic inflation, where the noise arises from the short wavelengths of some quantum field within *one and the same* system. For the case under consideration, the replica trick gives us a simple procedure to perform stochastic averages over the generating functional. There, each $\mathcal{Z}$ is a function of the random variables $\{h\}$, so its averaged n'th power $\mathcal{Z}^n$ is too. It is only the integration variables $\vec{\varphi}$ that acquires an additional label, namely the replica index $a = 1,\ldots,n$.

5.5 Variational Method

This section is devoted to the application of the Feynman-Jensen inequality and a Gaussian variational principle to derive variational equations, which allows us to obtain a closed expression for the full propagator of the long-wavelength modes.

5.5.1 Feynman-Jensen Inequality

To perform the stochastic average over the noise potential V, we use the replica trick and obtain

$$\overline{\mathcal{Z}^n} = \int \prod_{a=1}^{n}\mathcal{D}[\vec{\varphi}_a]\overline{\exp\left(-\sum_{b=1}^{n}\mathcal{S}[\vec{\varphi}_b]\right)} \equiv \int \prod_{a=1}^{n}\mathcal{D}[\vec{\varphi}_a]\exp\left(-\mathcal{S}^{(n)}[\vec{\varphi}]\right), \tag{5.21}$$

5.5 Variational Method

with the **replicated action**

$$\mathcal{S}^{(n)}[\vec{\varphi}] := \frac{1}{2} \sum_{a=1}^{n} \int_{t,t'} \int_{\boldsymbol{k}} G_0^{-1}(t,t',\boldsymbol{k}) \vec{\varphi}_a(t,\boldsymbol{k}) \cdot \vec{\varphi}_a(t',-\boldsymbol{k}) \\ - \frac{1}{2} \sum_{a,b=1}^{n} \int_{x,y} \phi(t,t') \delta(\boldsymbol{x}-\boldsymbol{y}) N \, \mathrm{R}\left(\frac{\vec{\varphi}_a(x) \cdot \vec{\varphi}_b(y)}{N}\right). \quad (5.22)$$

For simplicity, we used short-range product correlation and refrained from writing the different replica fields in the argument of the action, i.e.

$$[\vec{\varphi}] := [\vec{\varphi}_1, \ldots, \vec{\varphi}_n]. \quad (5.23)$$

We proceed with a **Feynman-Jensen variation principle** [Fey55] and therefore define the Gaussian trial action

$$\mathcal{S}_0[\vec{\varphi}] = \frac{1}{2} \sum_{a,b=1}^{n} \int_{t,t',\boldsymbol{k}} \mathrm{G}^{-1}{}_{ab}(t,t',\boldsymbol{k}) \vec{\varphi}_a(t,\boldsymbol{k}) \cdot \vec{\varphi}_b(t',-\boldsymbol{k}), \quad (5.24)$$

where we make the following ansatz for the propagator

$$\mathrm{G}^{-1}{}_{ab} := \left[\mathrm{G}_0^{-1} + \sigma_c + \sigma_{aa}\right]\delta_{ab} - \sigma_{ab}, \quad (5.25)$$

only making the replica matrix structure explicit.

Let us comment on the structure of (5.25). On the main diagonal (in replica space) we have the inverse of the noiseless propagator G_0^{-1} *plus* some mass correction σ_c, to be determined later, e.g. by the variational principle described below in this section. This alone would not only be trivial but also inconsistent, as we will see later. Hence, the off-diagonal part is filled by some, a priori unknown, **replica structure** σ_{ab}, which in general can be time dependent and, if one includes long-range noise correlation (see below), also momentum dependent, directly affecting the scaling behaviour of the propagator. Thus, although this variational method only generates a mass-energy contribution, its off-diagonal replica structure might have a viable influence on large-scale correlations.

The **Gaussian variational method** becomes exact in the limit $N \to \infty$ and allows one to go beyond ordinary perturbation theory. It is based on the **Feynman-Jensen inequality** [Fey55]

$$\ln(\mathcal{Z}) \geq \ln(\mathcal{Z}_0) + \left\langle \mathcal{S}_0^{(n)} - \mathcal{S}^{(n)} \right\rangle_0 =: \mathcal{F}_{\mathrm{var}}, \quad (5.26)$$

where the subscript 0 refers to the variational action (5.24) and we temporarily Wick-rotate to Euclidean signature. Equation (5.26) can easily be proven by using the Jensen inequality $\exp(\langle \ldots \rangle) \leq \langle \exp(\ldots) \rangle$, which comes from the convexity of the exponential. The problem is to find the best G_{ab}, i.e. the best σ_{ab}, by maximising the right-hand side of (5.26), which means to solve $\delta \mathcal{F}_{\mathrm{var}}/\delta G_{ab} = 0$.

Computing $\mathcal{F}_{\text{var}}$ per component and spatial volume yields

$$\begin{aligned}\frac{\mathcal{F}_{\text{var}}}{N\,\text{Vol}(d-1)} = &\frac{1}{2}\sum_{a=1}^{n}\int_{t,t',\mathbf{k}} G_0^{-1}(t,t',\mathbf{k})\,G_{aa}(t,t',\mathbf{k})\\ &-\frac{1}{2}\int_{t,t',\mathbf{k}}\text{Tr}\ln\bigl(G(t,t',\mathbf{k})\bigr) + C\\ &-\frac{1}{2}\sum_{a,b=1}^{n}\int_{t,t'}\phi(t,t')\widehat{R}\!\left(\int_{\mathbf{k}} G_{ab}(t,t',\mathbf{k})\right),\end{aligned}\qquad(5.27)$$

where we temporarily switched to finite spatial volume $\text{Vol}(d-1)$. The constant $C \in \mathbb{R}$, which vanishes after variation, includes $\langle \mathcal{S}_0 \rangle$ as well as terms from $\mathcal{F}_0$. For difference correlation (5.18b) one just has to replace the argument of $\widehat{R}$ by

$$\int_{\mathbf{k}}\bigl(G_{aa}(t,t',\mathbf{k}) + G_{bb}(t,t',\mathbf{k}) - 2G_{ab}(t,t',\mathbf{k})\bigr). \qquad(5.28)$$

The 'hat' over the function R in (5.28) is defined through

$$\widehat{R}(\langle\,\cdot\,\rangle_0) := \langle R(\,\cdot\,)\rangle_0 . \qquad(5.29)$$

In the limit $N \to \infty$, averaging and applying the (analytic) function R commute and so we drop the 'hat' when such a limit is considered.

The variation of $\mathcal{F}_{\text{var}}$, given in (5.28), with respect to the n^2 variational parameters G_{ab} gives for $a \neq b$:

$$\sigma_{ab}(t) = \phi(t)\widehat{R}'\!\left(\int_{\mathbf{k}} G_{ab}(t,\mathbf{k})\right), \qquad(5.30a)$$

$$\sigma_c(t) = -\phi(t)\widehat{R}'\!\left(\int_{\mathbf{k}} G_{aa}(t,\mathbf{k})\right), \qquad(5.30b)$$

where here a primed function denotes its differentiation with respect to its argument. We use

$$\frac{\delta G_{ab}(t,\mathbf{k})}{\delta G_{cd}(r,\mathbf{p})} = \delta(t-r)\,\delta^{(d-1)}(\mathbf{k}-\mathbf{p})\,\delta_{ac}\delta_{bd} \qquad(5.31)$$

and define for equal times $s_{ab}(t) := s_{ab}(t,t)$, $s_c(t) := s_c(t,t)$, $G_{ab}(t,\mathbf{k}) := G_{ab}(t,t,\mathbf{k})$, and $G_0(t,\mathbf{k}) := G_0(t,t,\mathbf{k})$.

Again, for difference correlation we obtain a similar result as (5.30b) (c.f. equation (3.12) in [MP91]) with the argument of $\widehat{R}'$ replaced by

$$\int_{\mathbf{k}}\bigl(G_{aa}(t,\mathbf{k}) + G_{bb}(t,\mathbf{k}) - 2G_{ab}(t,\mathbf{k})\bigr) \qquad(5.32a)$$

5.5 Variational Method

plus a global minus sign in front of σ. Additionally, we find in this case

$$\sigma_c(t) = -\sum_{\substack{a=1\\a\neq b}}^{n} \sigma_{ab}(t) \,. \tag{5.32b}$$

It is important to note that *any* interaction not containing noise just modifies the mass through its diagonal structure in replica space and that the free Gaussian case studied in the main text, *necessarily* yields replica symmetry.

The physical interpretation of the saddle-point equations (5.25), (5.30a,b), (5.32a,b) is as follows: The replica structure σ might be viewed as a generalised self energy. This follows from an expansion of the stationarity equations (5.30a,b) and (5.32a,b) in powers of (G_{ab}).

5.5.2 Long-Range Correlation

We now consider the case where the noise-correlation is non-local. This is described by the correlation

$$\overline{V(\bar{\varphi}_a(x),x)\,V(\bar{\varphi}_b(y),y)} = \phi(t,t',\boldsymbol{x}-\boldsymbol{y})\,N\,\mathrm{R}\!\left(\frac{\Xi_{ab}(x,y)}{N}\right). \tag{5.33}$$

Going through analogous steps as in the previous section, we obtain pendants to (5.30a,b),

$$\sigma_{ab}(t,\boldsymbol{p}) = \int_{\boldsymbol{x}} \phi(t,\boldsymbol{x})\,\mathrm{e}^{-\mathrm{i}\boldsymbol{p}\cdot\boldsymbol{x}}\,\widehat{\mathrm{R}}'\!\left(\int_{\boldsymbol{k}} \mathrm{e}^{-\mathrm{i}\boldsymbol{k}\cdot\boldsymbol{x}}\,G_{ab}(t,\boldsymbol{k})\right), \tag{5.34a}$$

$$\sigma_c(t,\boldsymbol{p}) = -\int_{\boldsymbol{x}} \phi(t,\boldsymbol{x})\,\mathrm{e}^{-\mathrm{i}\boldsymbol{p}\cdot\boldsymbol{x}}\,\widehat{\mathrm{R}}'\!\left(\int_{\boldsymbol{k}} \mathrm{e}^{-\mathrm{i}\boldsymbol{k}\cdot\boldsymbol{x}}\,G_{aa}(t,\boldsymbol{k})\right)$$
$$+ 2\widehat{\mathrm{U}}'\!\left(\int_{\boldsymbol{k}} G_{aa}(t,\boldsymbol{k})\right), \tag{5.34b}$$

for product correlation and to (5.32a,b),

$$\sigma_{ab}(t,\boldsymbol{p}) = -2\int_{\boldsymbol{x}} \phi(t,\boldsymbol{x})\,\mathrm{e}^{-\mathrm{i}\boldsymbol{p}\cdot\boldsymbol{x}}$$
$$\times \widehat{\mathrm{R}}'\!\left(\int_{\boldsymbol{k}} \mathrm{e}^{-\mathrm{i}\boldsymbol{k}\cdot\boldsymbol{x}}\,(G_{aa}(t,\boldsymbol{k}) + G_{bb}(t,\boldsymbol{k}) - 2G_{ab}(t,\boldsymbol{k}))\right), \tag{5.34c}$$

$$\sigma_c(t,\boldsymbol{p}) = -\sum_{\substack{a=1\\a\neq b}}^{n} \sigma_{ab}(t,\boldsymbol{p})\,, \tag{5.34d}$$

for difference correlation, where $a \neq b$.

We also generalised to a non-random potential U (with suitable rescaling by factors of N), which is so far taken to be arbitrary. It may arise from non-zero averages of the random distributions. The last equation is restricted to $\widehat{\mathrm{U}}' \equiv 0$.

Equations (5.34a-d) show that the replica matrix is in general space-time dependent, which affects the scaling behaviour of the two-point function (c.f. [KMT83]). Clearly, the short-range variational equations are included in (5.34a-d).

5.5.3 Replica Symmetric Propagator

In the previous section we derived the variational equations (5.34a-d) for a general matrix (σ_{ab}). It is important to try the simplest ansatz, which consists of taking $\sigma_{ab} = \sigma$ for all $a \neq b$, meaning that different replicas couple all in the same way among each other. This **replica symmetric** case is exact for quadratic as well as for quartic self-interactions. One finds

$$(G_{ab})^{-1} = \left[G_0^{-1} + \sigma_c\right]\mathbb{1} - \sigma(\mathbb{1} - 1) = G_0^{-1}\mathbb{1} - \sigma\mathbb{1}. \tag{5.35}$$

Thus, the inverse G_{ab} has the form

$$(G_{ab}) = G_0\mathbb{1} + \sigma G_0^2 \mathbb{1}, \tag{5.36}$$

where the physical limit $n \to 0$ has been taken and we define the $n \times n$-matrix $\mathbb{1}$ by $\mathbb{1}_{ij} = 1$ for all i, j. It has the property $\mathbb{1}^2 = n\mathbb{1} \xrightarrow{n \to 0} 0$. We observe that the limit of vanishing correlation, i.e. $\sigma \to 0$, gives back the free propagator as expected. The *physical* propagator $G(k)$ is given by

$$G(k) = \lim_{n \to 0} \frac{1}{n} \text{Tr}\left[(G_{ab}(k))\right]. \tag{5.37}$$

Chapter 6

Stochastic Inflation
&
Replica Field Theory

»μια υπέροχη αρμονία προκύπτει από συνδυασμό που συνδέει
φαινομενικά άσχετα.«

[»A wonderful harmony arises from joining together the seemingly unconnected.«]

according to HERACLITUS OF EPHESUS

This chapter brings together now what has been introduced in chapter 4 and 5, namely stochastic inflation and replica field theory. While the former treats quantum fluctuations in a stochastic manner, and the latter is designed for the treatment of a wide class of systems with a random potential, it seems tempting to combine them. By doing so, we study in a new non-perturbative way the dependence of the power spectrum on filter functions. If these deviate on an unbounded interval from the step function, a variant of the so-called dimensional reduction [AIM76, PS79, KLP84] is found, which results in a strong deviation from scale invariance in the infra-red, signalling a breakdown of the test-field assumption [KS08, KS09]. However, we show that scale invariance is preserved on all scales at late times for filters with bounded support (for wavelengths above the cut). Besides discussing the aspect of filter functions, we are able to treat general self-interactions. This is illustrated for the case of a quartic potential. Consequently we touch the important issue of whether our results for stochastic inflation might have an observable effect. We discuss how the cosmic microwave background radiation could be influenced.

In detail, this chapter is organised as follows: After calculating the free, noiseless propagator for exponential as well as for power-law inflation, we derive an expression for the effective infra-red power spectrum in section 6.2. Special focus is put on the dependence of the filter functions (section 6.3). A non-linearity parameter g_{NL} is introduced to quantify the modification of the power spectrum.

Then in section 6.4, we discuss the inclusion of self-interactions, where particular emphasis is put on the case of a quartic potential. This is followed by a discussion on possible effects of the obtained results on the cosmic microwave background radiation (section 6.5).

6.1 Free Power Spectrum

We devote this section to review the quantisation of a free, M-component, minimally-coupled, real, scalar test field $\vec{\Phi}$ with mass μ. We solve the classical mode equations, calculate the power spectrum and give expressions for the spectral index.

Let us concentrate on a spatially-flat, isotropic and homogeneous Universe in four-dimensional space-time. For its scale factor we will assume either exponential inflation, $a(t) := \mathrm{e}^{Ht}$, or power-law inflation, $a(t) := (t/t_1)^p$ with $p > 1$ and the reference time t_1 defined by $a_1 := a(t_1) = 1$. For convenience we use $\hbar \stackrel{!}{=} c \stackrel{!}{=} 1$ and set $H \stackrel{!}{=} 1$ in the case of de Sitter or $t_1 \stackrel{!}{=} 1$ for power-law inflation. The mode function $u(t, k)$ is defined by the decomposition of the field components

$$\Phi_i(t, \mathbf{k}) = \hat{a}_i(\mathbf{k}) u(t, k) + \text{H.c.} , \qquad (6.1)$$

with the modulus of the comoving momentum $k := |\mathbf{k}|$. The annihilation and creation operators obey the commutation relations

$$\left[\hat{a}_i(\mathbf{p}), \hat{a}_j^\dagger(\mathbf{k})\right] = (2\pi)^3 \delta^3(\mathbf{p} - \mathbf{k}) \delta_{ij} , \qquad (6.2a)$$

$$\left[\hat{a}_i(\mathbf{p}), \hat{a}_j(\mathbf{k})\right] = 0 . \qquad (6.2b)$$

The rescaled mode functions $v(\eta, k) := a(\eta) u(\eta, k)$ fulfil the mode equation

$$v'' + \left[k^2 + \mu^2 a^2 - \frac{a''}{a}\right] v = 0 , \qquad (6.3)$$

with primes denoting derivatives with respect to conformal time η. Solutions to (6.3) are fixed by requiring that for very short wavelengths the effect of space-time curvature and mass becomes irrelevant, and thus a Minkowski solution should be obtained, i.e.

$$\lim_{\frac{k}{a} \to \infty} v(\eta, k) = \frac{\mathrm{e}^{-\mathrm{i}k\eta}}{\sqrt{2k}} . \qquad (6.4)$$

The factor $1/\sqrt{2k}$ is fixed by the canonical commutation relations of φ and its conjugate momentum.

Recall that exponential inflation implies $a(\eta) = |\eta|^{-1}$ and for power-law models one finds $a(\eta) \propto |\eta|^{-\frac{p}{p-1}}$, where both cases match in the limit $p \to \infty$. Then, equation (6.3) can be rewritten in the

6.1 Free Power Spectrum

form

$$v'' + \left[k^2 - \frac{1}{\eta^2}\left(\nu^2 - \frac{1}{4}\right)\right] v = 0 \,, \tag{6.5}$$

with

$$\nu = \sqrt{\frac{9}{4} - \mu^2} \qquad \text{(exponential)}\,, \tag{6.6a}$$

and for zero mass

$$\nu = \frac{p}{p-1} + \frac{1}{2} \qquad \text{(power-law)}\,. \tag{6.6b}$$

A general solution to (6.5), fulfilling (6.4), is given in terms of Bessel functions:

$$u(\eta, k) = \frac{\sqrt{\pi}}{2} \frac{\sqrt{|\eta|}}{a(\eta)} \left[J_\nu(k\,|\eta|) + \mathrm{i}\, Y_\nu(k\,|\eta|) \right]. \tag{6.7}$$

On large scales (for $k \to 0$) and for positive ν, the leading term of (6.7) is

$$u(\eta, k) \simeq -\frac{\mathrm{i}\, 2^{\nu-1}\, |\eta|^{1/2-\nu}\, \Gamma(\nu)}{\sqrt{\pi}\, a(\eta)}\, k^{-\nu}\,, \tag{6.8a}$$

while for negative ν one has

$$u(\eta, k) \simeq \frac{2^{-\nu-1}\, \sqrt{\pi}\, |\eta|^{1/2+\nu}\, \left(\mathrm{i}\cot(\pi\nu) + 1\right)}{\Gamma(\nu+1)\, a(\eta)}\, k^\nu\,. \tag{6.8b}$$

The free (equal-time) propagator $G_0(t, \boldsymbol{p}, \boldsymbol{k})$ shall be defined via

$$G_0(t, \boldsymbol{p}, \boldsymbol{k})\,(2\pi)^3\,\delta^3(\boldsymbol{k} - \boldsymbol{p}) \equiv \frac{1}{2M} \left\langle \Omega \left| \vec{\Phi}(t, \boldsymbol{p}) \cdot \vec{\Phi}^\dagger(t, \boldsymbol{k}) \right| \Omega \right\rangle, \tag{6.9}$$

where the vacuum $|\Omega\rangle$ is defined by $\hat{a}(k)|\Omega\rangle = 0$ and a subscript '0' indicates a quantity that is calculated in the absence of any classical noise.

An object of central interest in cosmology is the dimensionless **power spectrum** $\mathcal{P}(t, k)$. As explained previously, its relation to some (equal-time) propagator $\mathrm{G}(t, k) := \mathrm{G}(t, \boldsymbol{k}, \boldsymbol{k})$, with assumed infra-red behaviour $\mathrm{G}(t, k) \sim k^{-\varkappa}$, is given by

$$\mathcal{P}(k) := k^3\, \mathrm{G}(k) \sim k^{n_s - 1}\,, \tag{6.10}$$

where we omitted the time argument for the sake of brevity. The **spectral index** n_s is connected to the so-called **critical exponent** $\varkappa$ via $n_s = 4 - \varkappa$. For $\mu = 0$ the power spectrum of the free, noiseless theory is scale invariant, i.e. $n_s = 1$ for $\varkappa_0 = 3$. Non-zero mass leads to

$$n_s = 4 - 3\sqrt{1 - \frac{4}{9}\mu^2} = 1 + \frac{2}{3}\mu^2 + \mathcal{O}(\mu^4)\,, \tag{6.11a}$$

and the power-law case with $\mu = 0$ yields

$$n_s = \frac{p-3}{p-1} = 1 - \frac{2}{p} + \mathcal{O}\left(\frac{1}{p^2}\right). \tag{6.11b}$$

Note that $n_s(\mu \neq 0) > 1$ for exponential inflation, while $n_s < 1$ in the massless power-law case. A scale-invariant spectrum is recovered in the limits $\mu \to 0$ or $p \to \infty$, respectively. Also note that the results (6.11a) and (6.11b) do not include any metric perturbation, which would generally cause negative deviations from $n_s = 1$ (c.f. [LLM+02]).

6.2 Effective Power Spectrum

This section contains the study of the infra-red behaviour of the physical propagator G and therefore of the power spectrum $\mathcal{P}$. We first study some rather general set-up and discuss the phenomenon of dimensional reduction. Then we focus on the specific case of stochastic inflation.

6.2.1 Long-Range Correlation

The full two-point function for the long-wavelength modes, $\mathrm{G} := \frac{1}{M}\overline{\langle \vec{\varphi} \cdot \vec{\varphi} \rangle}$, has been calculated in the previous chapter, using a Feynman-Jensen-type variational calculation together with replica field theory. From equation (5.36) we have

$$\mathrm{G}(t,k) = \mathrm{G}_0(t,k) + \sigma(t,k)\,\mathrm{G}_0(t,k)^2 \,, \tag{6.12}$$

where the replica structure σ is determined by the variational equations (5.34a,b).

To analyse the physical consequence of equation (6.12) on the power spectrum, let us now assume for the two-point noise correlation $\overline{\mathrm{h}_i(\boldsymbol{x})\mathrm{h}_j(\boldsymbol{y})} = \phi(\boldsymbol{x}-\boldsymbol{y}) \sim |\boldsymbol{x}-\boldsymbol{y}|^{-3+\rho}$ with $\rho < 3$. This is a case of the so-called long-range correlation (c.f. section 5.5.2), which describes properly the infra-red limit of the physical model discussed below. In momentum space, the above choice implies $\phi(k) \sim k^{-\rho}$ and hence $\sigma(t,k) = \wp(t)k^{-\rho}$ by virtue of (5.34a,b). For $\rho > -\varkappa_0$, the infra-red behaviour of the power spectrum deviates from the noiseless result, and we find $\varkappa = 2\varkappa_0 + \rho$. This result is consistent with previous studies in flat space for a propagator with $\varkappa_0 = 2$ [KMT83].

What does this mean for the spectral index? Since $n_s = 4 - \varkappa$, we find the result

$$n_s = 4 - 2\varkappa_0 - \rho \tag{6.13}$$

if $\rho > -\varkappa_0$. Thus, one would find a dramatic change of the super-horizon power spectrum as compared to the case without noise. Specifically, for exponential inflation this implies a modification of the spectral index on large super-horizon scales if the spatial noise correlator decreases at most like

6.2 Effective Power Spectrum

$|x|^{-6}$ if $\mu = 0$, while for finite mass this exponent changes to $-6 + 2/3\mu^2 + \mathcal{O}(\mu^4)$. In the massless power-law model the power is given by $-6 - 2p^{-1} + \mathcal{O}(p^{-2})$. Equation (6.13) constitutes one variant of the phenomenon of **dimensional reduction** [AIM76, PS79, KLP84], which can rigorously be proven to all orders in perturbation theory for $\rho = 0$ and for arbitrary non-random potentials (see especially [KLP84] for a supersymmetric version of the proof). Because this effect originates from the second piece of G in (6.12) it will be referred to as the **dimensional reduction part**.

Please note that those changes only concern the power spectrum of the smoothed (classical) long-wavelength modes, which are influenced by their short (quantum) counterparts. It does not mean that the full quantum two-point function obeys dimensional reduction. We should underline that the above statements on dimensional reduction depend on the specific choice of the filter function. Their influence on the power spectrum is discussed in section 6.3.

A natural question to ask is on which scales the effect of dimensional reduction shows up. Let us therefore define the **transition scale** k_* at which the two terms on the right-hand side of equation (6.12) balance each other via

$$G_0(t, k_*) \stackrel{!}{=} \sigma(t, k_*)\, G_0(t, k_*)^2 \,. \tag{6.14}$$

It separates two regions such that for $k \gg k_*$ the behaviour is noiseless and for $k \ll k_*$ dimensional reduction holds.

6.2.2 Stochastic Inflation

Let us now return to our physical model of stochastic inflation. The split of the field $\vec{\Phi}$ into a long- and short-wavelength part, $\vec{\Phi} = \vec{\varphi} + \vec{\phi}$, together with the free field equation $(\Box + \mu^2)\vec{\Phi} = \vec{0}$ implies for the infra-red part of the propagator

$$G(t, k) \simeq \left| \left(\tilde{\Box}_k + \mu^2 \right) \left[W_\kappa \left(\frac{k}{a(t)} - \epsilon \right) u(t, k) \right] \right|^2 |u(t, k)|^4 , \tag{6.15}$$

where $\tilde{\Box}_k$ is the (spatially Fourier-transformed) covariant Laplacian, $u(t, k)$ is the mode function from equation (6.1), and W_κ is a smooth high-pass filter (c.f. section 6.3), cutting out the large wavelengths above ϵ. The parameter κ controls the width of the cut. In the limit $\kappa \to 0$, W_κ approaches a step function. Here we choose

$$W_\kappa(z) = \frac{1}{\pi} \arctan\left(\frac{z}{\kappa}\right) + \frac{1}{2} \tag{6.16}$$

and take $0 < \epsilon \ll 1$ in order to separate at wavelengths well below the Hubble scale $H^{-1}(=1)$, and $\kappa \ll \epsilon$ to have a narrow transition region between quantum and classical modes. We do not impose any restriction on μ except that we demand the radicand in (6.11a) to be positive, i.e. $\mu^2 \leq 9/4$.

Please note that the filter function (6.16) does not have a bounded support for wavelengths above the cut scale, meaning that also modes from the far infra-red influence the quantum noise. A further discussion on filter functions, in particular of such with bounded support (above the cut), is presented in the subsequent section.

Using (6.16), one finds that the model given in (6.15) is of long-range-type and implies

$$\rho = 3\sqrt{1 - \frac{4}{9}\mu^2} - 2 = 1 - \frac{2}{3}\mu^2 + \mathcal{O}(\mu^4) \qquad (6.17a)$$

for exponential inflation, and

$$\rho = \frac{p+1}{p-1} = 1 + \frac{2}{p} + \mathcal{O}\left(\frac{1}{p^2}\right) \qquad (6.17b)$$

for massless power-law models. Thus for $k \ll k_*$ we obtain with (6.13)

$$n_s = 6 - 9\sqrt{1 - \frac{4}{9}\mu^2} = -3 + 2\mu^2 + \mathcal{O}(\mu^4) \qquad (6.18a)$$

in the exponential case, and

$$n_s = 3\frac{p+1}{1-p} = -3 - \frac{6}{p} + \mathcal{O}\left(\frac{1}{p^2}\right) \qquad (6.18b)$$

for power-law inflation with $\mu = 0$. In the infra-red limit we see that scale invariance of the effective power spectrum is destroyed even in the massless, exponential inflationary scenario. Also the power-law case changes drastically.

For scales $k \gg k_*$ the noiseless spectral index (6.11a) is recovered. The late-time behaviour of the transition scale k_*, defined in equation (6.14), can be calculated analytically:

$$k_* = \left(e^{-t}\right)^{\frac{8-2x}{2x-2}} \pi^{-\frac{2}{x-1}} \left(\frac{2^{2x-3}(5 - 2\mu^2 - x)\kappa^2 \Gamma\left(\frac{1}{2}x\right)^4}{(\epsilon^2 + \kappa^2)^2}\right)^{\frac{1}{2x-2}}, \qquad (6.19)$$

where

$$x := \sqrt{9 - 4\mu^2}, \qquad (6.20)$$

for exponential inflation, and

6.2 Effective Power Spectrum

$$k_* = 2\left(\frac{p}{t^{p+1}}\right)^{\frac{1}{2}-\frac{1}{2p}} \left(\frac{\kappa \Gamma\left(\frac{3}{2}+\frac{1}{p-1}\right)^2}{\pi^2 \left(\epsilon^2 + \kappa^2\right)}\right)^{\frac{1}{2}-\frac{1}{2p}} \quad (6.21)$$

for massless power-law inflation. In the zero-mass limit (6.19) yields the asymptotic form

$$k_*(t) = e^{-t/2} \frac{2\sqrt{\kappa}}{\sqrt{\pi}\sqrt{\epsilon^2 + \kappa^2}}. \quad (6.22)$$

Thus for $\epsilon \neq 0$, k_* goes to zero in the (step-function) limit $\kappa \to 0$, i.e. dimensional reduction is absent — a statement that holds for both exponential and power-law inflation. This is a general feature of the free theory where is no mixing of the short (quantum) modes with the long classical ones as a sharp cut-off is introduced. In a slightly different setup, with a filter function having only one parameter, i.e. $\epsilon = \kappa$, this has already been noted in reference [MMR04].

Let us turn to the issue of the compatibility of (6.21) with (6.19). As has already been mentioned in section 6.1, for $p \to \infty$ and $\mu = 0$, both cases should match. A naive limit of (6.21) shows that this is not obvious. The point here is that one should carefully look at the time dependence. Expressing (6.21) in terms of the number of e-folds, $N := \ln(a/a_1)$, shows indeed the desired coincidence. So for all plots related to power-law cases, we will make the replacement

$$t = \left(p e^{\tilde{t}}\right)^{\frac{1}{p+1}}. \quad (6.23)$$

A straightforward calculation shows the relation between $\tilde{t}$ and N,

$$N = \frac{p}{p+1}\left[\tilde{t} + \ln(p)\right] = \tilde{t} + \ln(p) + \mathcal{O}\left(p^{-1}\right). \quad (6.24)$$

This means that for large p, $\tilde{t} = N$ up to a shift which originates from different time normalisation of the power-law and the de Sitter case where $t = N$. This shift is indeed visible in figures 6.4 and 6.6, where only the last part of the transient phenomenon shows up, contrary to the corresponding exponential-inflation plots, where a larger part can be observed (details below).

In figure 6.1 (figure 6.2) we show the effective long-wavelength power spectrum $\mathcal{P}$ as a function of k for fixed time $Ht = 10$ and mass $\mu = 0.1$ ($\tilde{t} = 4$ and $p = 12$) with $\epsilon = 10^{-2}$ and $\kappa = 10^{-3}$ for the de Sitter (massless power-law) model. One can see that it diverges stronger than the noiseless power spectrum as k tends to zero, putting therefore more correlation on large scales. Furthermore one sees that the part $k^3 \sigma G_0^2$ approximates the full power spectrum $\mathcal{P}$ in the infra-red as well as that the noiseless piece $k^3 G_0$ gives a suitable ultra-violet approximation. In all plots we took the filter function (6.16). Figure 6.3 shows the time behaviour of the comoving momentum k_* for exponential inflation for different values of the mass μ. Figure 6.4 displays the same for the power-law model. The solid rays represent the analytic approximations (6.19) and (6.21), respectively, while the dashed curves are the full results, obtained numerically from (6.14) using the FindRoot function of MATH-EMATICA 7. Well below this borderline the two-point function obeys dimensional reduction, while well above ordinary scaling holds. After an initial transient phenomenon, whose duration depends on the specific choice of ϵ and κ, the comoving transition scale decays exponentially fast. Hence, the dimensional-reduction contribution is pushed to larger and larger scales as time increases. This therefore guarantees that quantum modes induce only a minor change of the spectral index on sub-horizon as well as on moderate super-horizon scales.

For the sake of being specific, let us consider a mode with comoving momentum $k = 0.05\,H$. At $t = 0$ it is within the region of ordinary scaling, suffering at most slightly from dimensional reduction. This mode enters then, after roughly two e-folds, the region of broken scale invariance, but leaves it at the latest (for $\mu = 0$) after seven e-folds and stays eternally in the scale-invariant regime, which itself grows exponentially fast.

6.3 Filter Functions

The discussion of possible smooth filter functions and their influence on the phenomenon of dimensional reduction shall be dealt with in this subsection. In particular, we study their effects on the transition scale $k_*(t)$.

Let Θ be the Heaviside function. Then we define a **filter function** as a function W_κ depending on a parameter κ (controlling the width of the transition) such that

$$\lim_{\kappa \to 0} W_\kappa(z) \equiv \Theta(z). \qquad (6.25)$$

One may divide filter functions fulfilling (6.25) into two classes: Those for which $\Theta - W_\kappa$ has an unbounded support J and those for which J is bounded. Let us now discuss these two cases separately.

Unbounded Support

Some well-known smooth filter functions are:

$$\frac{\tan^{-1}\left(\frac{z}{\kappa}\right)}{\pi} + \frac{1}{2}, \qquad (6.26a)$$

$$\frac{1}{2}\operatorname{erf}\left(\frac{z}{\kappa}\right) + \frac{1}{2}, \qquad (6.26b)$$

$$\frac{\operatorname{Si}\left(\frac{\pi z}{\kappa}\right)}{\pi} + \frac{1}{2}, \qquad (6.26c)$$

6.3 Filter Functions

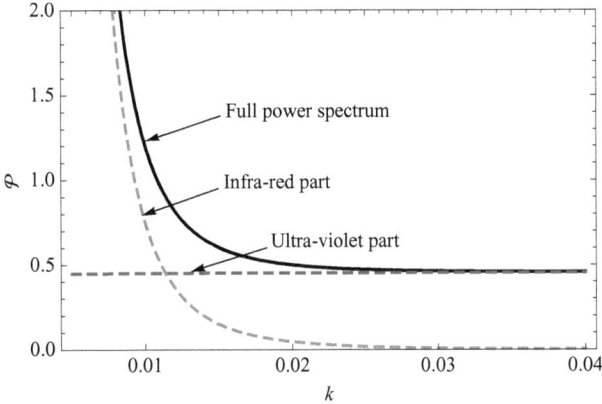

Figure 6.1: Power spectrum $\mathcal{P}$ for a massive test field ($\mu = 0.1H$) for exponential inflation at $tH = 10$ as a function of comoving momentum k. The parameters of the filter function (6.16) are fixed to $\kappa = 10^{-3}$ and $\epsilon = 10^{-2}$.

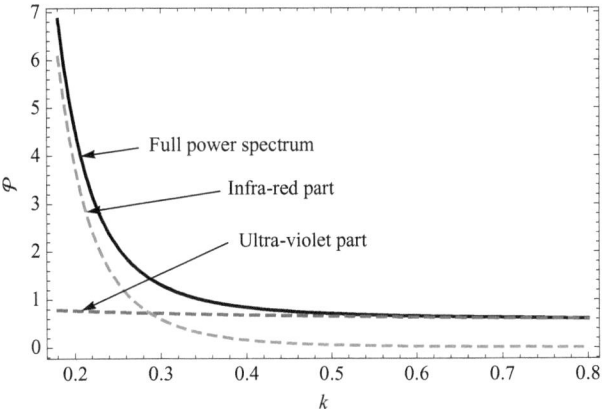

Figure 6.2: Power spectrum $\mathcal{P}$ for a massless test field for power-law inflation ($p = 12$) at $\tilde{t} = 4$ [c.f. (6.23)] as a function of comoving momentum k (in units of $1/t_1$). The parameters of the filter function (6.16) are fixed as in figure 6.1.

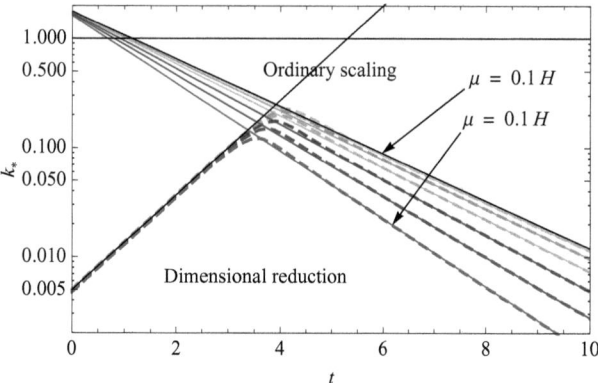

Figure 6.3: Comoving transition scale k_* [c.f. (6.19)] for exponential inflation as a function of cosmic time t (in units of H^{-1}) for mass $\mu/H = 0.1, 0.2, 0.3, 0.4, 0.5, 0.6$ (dashed lines, top to bottom). Dashed curves are numerical results, coloured solid lines are analytic approximations, and enveloping black lines are $\frac{\epsilon}{2}a(t)$ and the asymptotic form (6.22), respectively. The parameters of the filter function (6.16) are fixed as in figure 6.1.

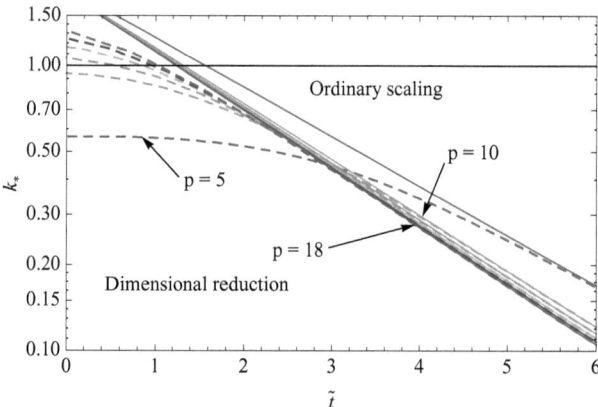

Figure 6.4: Comoving transition scale k_* for power-law inflation as a function of modified cosmic time $\tilde{t}$ [c.f. (6.23)] for $p = 5, 10, 12, 14, 16, 18$ (dashed lines, top to bottom). Dashed curves are numerical results, coloured solid lines are analytic approximations. The parameters of the filter function (6.16) are fixed as in figure 6.1.

6.3 Filter Functions

$$\frac{1}{1+e^{-\frac{z}{\kappa}}}, \tag{6.26d}$$

$$e^{-e^{-\frac{z}{\kappa}}}, \tag{6.26e}$$

$$\frac{1}{2}\tanh\left(\frac{z}{\kappa}\right) + \frac{1}{2}. \tag{6.26f}$$

Here, 'erf' is the error function, defined by $\mathrm{erf}(z) := \frac{2}{\sqrt{\pi}} \int_0^z \mathrm{d}t\ e^{-t^2}$, and 'Si' is the sine-integral function, defined by $\mathrm{Si}(z) := \int_0^z \mathrm{d}t\ \sin(t)/t$.

The following two figures show the dependence of the transition scale on various filter functions, where we choose the functions (6.26a), (6.26d) and (6.26f). It can clearly be seen that only the quantitative behaviour changes, i.e. the position of the 'bump', which marks the end of the transient phenomenon, and *not* the qualitative shape. Note that the curves have been rescaled by a fixed factor (one for each filter function) and that those functions all have precisely the same asymptotic behaviour. These statements concern both the exponential as well as the power-law case.

In figure 6.7 we display the dependence of k_* on various values of the width parameter κ. We find that decreasing this parameter shifts the curves downwards, therefore pushing the dimensional reduction effect to larger and larger scales. This can also be seen directly from the late-time formulae (6.19). The power-law case behaves similarly.

Bounded Support

We note that all of the filter functions (6.26a-f) do *not* have a lower bound on their support — a crucial ingredient for the occurrence of dimensional reduction in the far infra-red. This can be seen as follows: From equation (6.12) one observes that the dimensional-reduction part σG_0^2 is proportional to σ, which itself is related to the filter function W_κ in such a way that if $W_\kappa \neq \Theta$ on an interval J only, $\sigma = 0$ outside of J. Consequently, $G = G_0$, i.e. dimensional reduction is absent, on the complement $\mathbb{R}\backslash J$.

However, this does not mean that one can forget about dimensional reduction in the context of stochastic inflation: *Any* smooth filter function W_κ will definitely cause a deviation from scale invariance, although it might be that this deviation disappears for scales outside the support of $W_\kappa - \Theta$. Furthermore, the results of this work are not arbitrary, since an ultimate derivation of the stochastic inflation paradigm from first principles will single out a specific filter function.

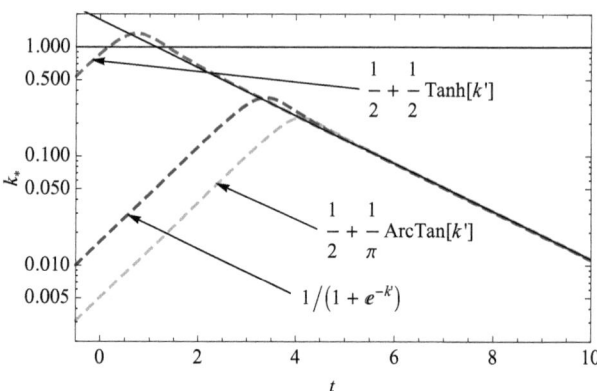

Figure 6.5: Influence of filter functions on the comoving transition scale k_* for exponential inflation as a function of cosmic time t (in units of H^{-1}) for mass $\mu/H = 0.1$. A smoothing width $\kappa = 10^{-3}$ and short-wavelength cut $\epsilon = 10^{-2}$ are chosen. The variable k' is a short-hand notation for $(k\eta - \epsilon)/\kappa$.

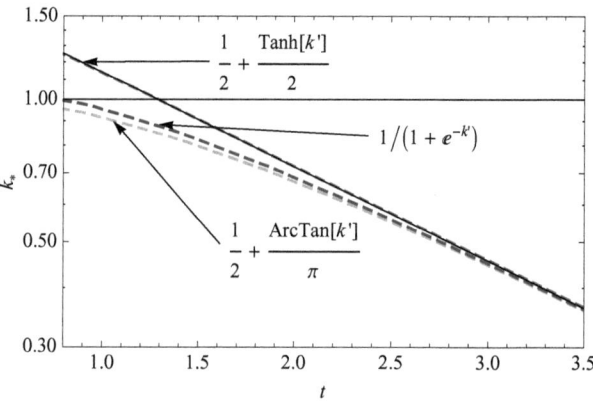

Figure 6.6: Influence of filter functions on the comoving transition scale k_* for power-law inflation as a function of modified cosmic time $\tilde{t}$ [c.f. (6.23)] for $p = 12$. Filter argument and parameters are fixed as in figure 6.5.

6.3 Filter Functions

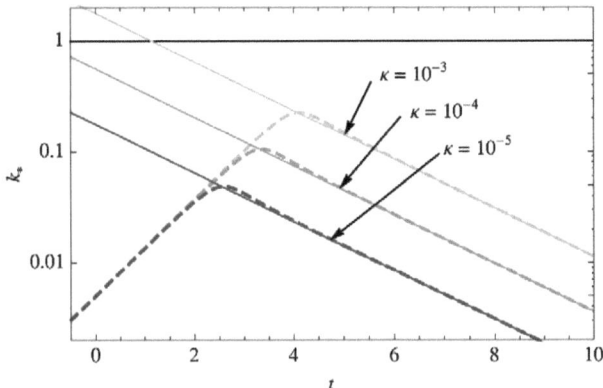

Figure 6.7: Dependence of the comoving transition scale k_* on the width parameter κ of filter (6.16) for exponential inflation as a function of cosmic time t (in units of H^{-1}) for mass $\mu/H = 0.1$ and $\epsilon = 10^{-2}$.

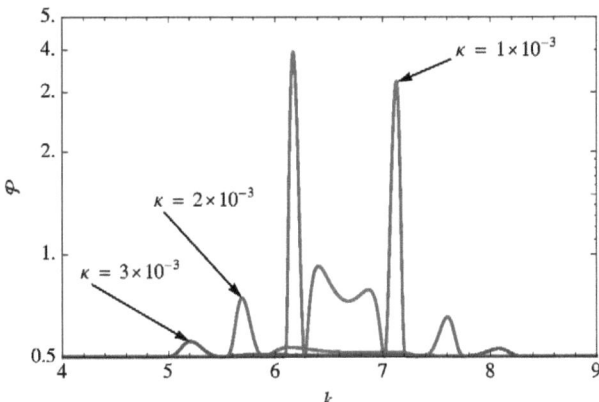

Figure 6.8: Effective power spectrum $\mathcal{P}(k)$ for filter (6.28) with bounded interval $\mathsf{J} =]-\kappa, +\kappa[$ for different values of κ. $\mathcal{P}(k)$ is evaluated for exponential expansion at $t = 6.5H$ with $\mu = 0$ and $\epsilon = 10^{-2}$.

We now study the effect of generic filter functions (with support that is bounded from below) on the power spectrum. For $J =]-\kappa, +\kappa[$ one may choose

$$W_\kappa(k') = \begin{cases} 0 & : \quad k' \leq -\kappa, \\ k' & : \quad k' \in J, \\ 1 & : \quad k' \geq +\kappa \end{cases} \tag{6.27a}$$

as a prototype filter function, where $k' := (k\eta - \epsilon)/\kappa$. Hence, its derivative W'_κ is given by

$$W'_\kappa(k') = \begin{cases} 0 & : \quad k' \leq -\kappa, \\ 1 & : \quad k' \in J, \\ 0 & : \quad k' \geq +\kappa, \end{cases} \tag{6.27b}$$

leading us to the smooth approximation

$$W'_\kappa(k') \simeq \begin{cases} 0 & : \quad k' \leq -\kappa, \\ \exp\left(1 - \frac{\kappa^4}{(\kappa^2 - k'^2)^2}\right) & : \quad k' \in J, \\ 0 & : \quad k' \geq +\kappa. \end{cases} \tag{6.28}$$

Figure 6.8 shows the influence of $J \neq \mathbb{R}$ on the power spectrum for the filter function due to (6.28). Although for $k\eta > \epsilon + \kappa$ the dimensional reduction effect disappears, one clearly has an effect inside the interval J. As the size of J shrinks, the domain in wave-number space for which the dimensional reduction part is dominant also sizes down, albeit the magnitude of $\mathcal{P}(k)$, as an effect of wavelength separation, increases considerably. This is reasonable, since the (step-function) limit $\kappa \to 0$ contains second derivatives on the Heaviside function, which correspond to the pole forming for (6.28).

In figure 6.9 we display the effective power spectrum $\mathcal{P}(k)$ at different times. As anticipated, the dimensional reduction 'bumps' decline as time increases. One also observes the same behaviour as in figure 6.3, namely the grows of the comoving transition scale, up to the point where the dimensional reduction effect disappears. It is subdominant in the shaded region of figure 6.10, which visualises the ratio

$$\eta := \frac{\sigma G_0^2}{G_0} = \sigma G_0 \tag{6.29}$$

of the dimensional reduction part to the noiseless part, evaluated at the most ultra-violet peak (c.f. figure 6.8).

One sees that η diminishes exponentially fast in time. Hence, after some few e-folds, the transition region becomes unimportant (shaded region in figure 6.10) and the classical power spectrum provides an excellent approximation.

6.3 Filter Functions

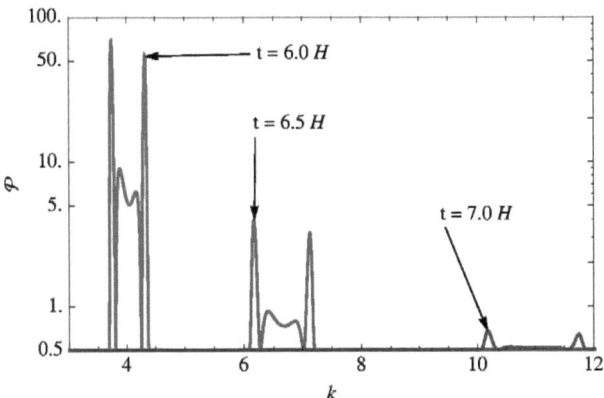

Figure 6.9: Effective power spectrum $\mathcal{P}(k)$ at various times and for $\kappa = 10^{-3}$, otherwise as in figure 6.8.

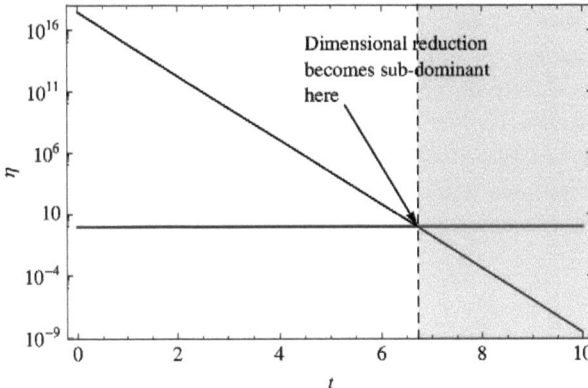

Figure 6.10: Time dependence of the ratio η of the dimensional reduction part to the noiseless part of the effective power spectrum for the filter (6.28) for exponential expansion with $\kappa = 10^{-3}$, $\epsilon = 10^{-2}$ and $\mu = 0$.

6.3.1 Modified Gaussian Fluctuations

One may connect the replica structure σ to a non-linearity parameter g_{NL}, which can in general be defined by

$$\varphi_i(t,k) \equiv \varphi_i^{\text{G}}(t,k) - g_{\text{NL}}(t,k)\left(\varphi_i^{\text{G}}(t,k)\right)^2, \tag{6.30}$$

where $\tilde{\varphi}^{\text{G}}(t,k)$ is a free Gaussian field. On the level of propagators, this translates to

$$G(t,k) = G_0(t,k) + 3g_{\text{NL}}(t,k)^2 G_0(t,k)^2 \tag{6.31}$$

and hence

$$\sigma(t,k) = 3g_{\text{NL}}(t,k)^2 \tag{6.32}$$

can be directly read off, using equation (6.12). The quantity g_{NL} measures the influence of the quantum fluctuations, picked up by a smooth filter function. Formally, it resembles an effective non-Gaussianity parameter [KS01] for the long-wavelength modes. However, this association is misleading since the full theory we started with is Gaussian (but with a non-trivial replica structure).

Unbounded Support

Let us first consider the case of a filter function for which J is unbounded. Exemplary we choose the function (6.16). Figure 6.11 shows the dependence of g_{NL} on the comoving momentum k for various values of μ for fixed time $t = 10H$, using equation (6.32). Firstly, one sees that increasing μ shifts the curve upwards, and secondly, one observers a divergence in the infra-red — displaying the effect of dimensional reduction. For $k \gg k_*$ one obtains a scale-invariant spectrum. Figure 6.12 visualises the same for power-law inflation with $\mu = 0$ for various values of p, where $\tilde{t}$ has been fixed to $\tilde{t} = 4$. One observes that increasing p lowers the curves which converge towards their asymptotic value for $p \to \infty$. This coincides with the $\mu \to 0$-limit of the exponential case as already noted in section 6.1.

Bounded Support

To discuss the effect of a bounded support J on the non-linearity parameter g_{NL}, we choose the filter corresponding to (6.28). The dependence of g_{NL} on the comoving momentum k is depicted in figure 6.13. All curves shown in this plot are strictly zero outside the plotted interval. This means that there is only a small ($\sim \kappa$) window around ϵ in which the wavelength-separation effects play a role at all. However, for sufficiently early times, the transition effect becomes indeed pronounced. Figure 6.14 shows the time dependence of the non-linearity parameter g_{NL} for various masses. After some few e-folds, g_{NL} is completely negligible and Gaussianity of the fluctuations in the proper sense holds true.

6.3 Filter Functions

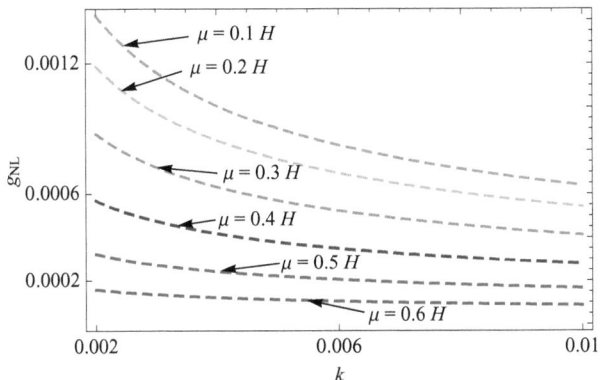

Figure 6.11: Non-linearity parameter g_{NL} for exponential inflation as a function of comoving momentum k (in units of H) for mass $\mu/H =$ 0.1 (uppermost), 0.2, 0.3, 0.4, 0.5 and 0.6 (lowermost) at $Ht = 10$. For the filter (6.16) we fix $\kappa = 10^{-3}$ and $\epsilon = 10^{-2}$.

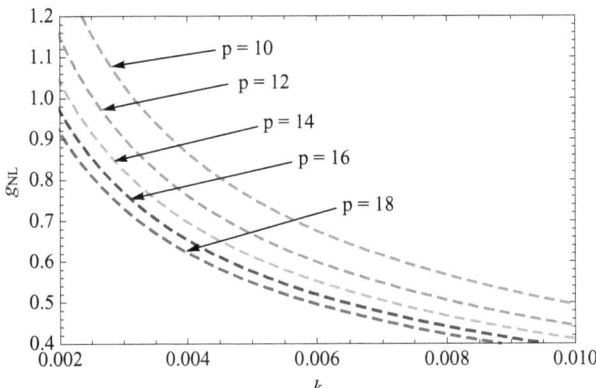

Figure 6.12: Non-linearity parameter g_{NL} for power-law inflation as a function of comoving momentum k (in units of $1/t_1$) for $p = 10$ (uppermost), $12, 14, 16, 18$ (lowermost) and $\tilde{t} = 4$. Filter and parameters are as in figure 6.11.

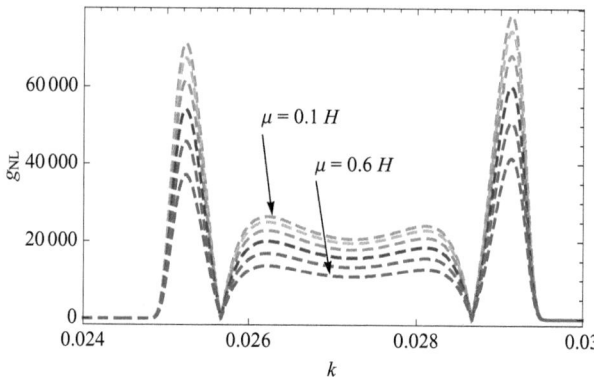

Figure 6.13: Non-linearity parameter g_{NL} as a function of comoving momentum k (in units of H) for de Sitter inflation with mass $\mu/H = 0.1$ (uppermost), 0.2, 0.3, 0.4, 0.5 and 0.6 (lowermost) with $Ht = 1$. The filter (6.28) is used with $\kappa = 10^{-3}$ and $\epsilon = 10^{-2}$.

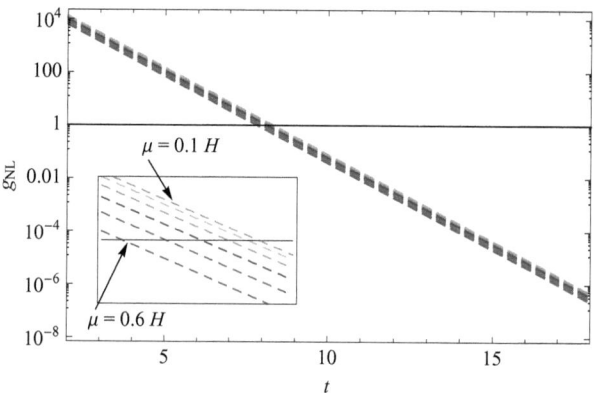

Figure 6.14: Non-linearity parameter g_{NL} as a function of time t (in units of H^{-1}) for de Sitter inflation with mass $\mu/H = 0.1, 0.2, 0.3, 0.4, 0.5, 0.6$ (dashed lines, top to bottom) evaluated at the first ultra-violet 'bump' of figure 6.13. Filter and parameters are as in figure 6.13.

6.4 Self-Interactions

In this section we extend the results of the previous sections, which have been published in [KS08, KS09], to include self-interactions of a massive, scalar test field in a de Sitter background. We put special emphasis on the calculation of the power spectrum and show that self-couplings cause a lack of power on large scales. This is explicitly demonstrated for the case of a quartic self-interaction.

Our starting point is the following Lagrangian for the massive scalar field Φ with a quartic self-interaction,

$$\mathcal{L} = \frac{1}{2} g^{\mu\nu} \partial_\mu \Phi \partial_\nu \Phi - \frac{\mu^2}{2} \Phi^2 - \frac{\lambda}{4} \Phi^4 , \tag{6.33}$$

where μ is the mass and λ is a self-coupling constant. Furthermore, we assume a de Sitter background geometry with an exponential scale factor $a(t) = \exp(t)$, where again, for convenience we use $\hbar \stackrel{!}{=} c \stackrel{!}{=} H \stackrel{!}{=} 1$. After the general study of chapter 5, we can easily write down the solution to the variational equations (5.34a,b). They are replica symmetric and read for $a \neq b$:

$$\sigma_{ab} = \sigma , \tag{6.34a}$$

$$\sigma_c = -\sigma + 4\lambda G_{aa} . \tag{6.34b}$$

Equations (6.34a) and (6.34b) imply the following self-consistent equation for the power spectrum

$$\mathcal{P}(k) = k^3 \left[G_0^{-1}(k) + 8\lambda k^{-3} \mathcal{P}(k) \right]^{-1} + k^3 \sigma(k) \left[G_0^{-1}(k) + 8\lambda k^{-3} \mathcal{P}(k) \right]^{-2} . \tag{6.35}$$

The explicit form of the solution to (6.35) for arbitrary values of μ and t is rather lengthy and shall not be given here. Instead we present its late-time expression for zero mass:

$$\mathcal{P}(k) \simeq \frac{\sqrt{4 + 6\lambda k^{-6}} \cos\left(\frac{1}{3} \arctan\left(\frac{6\sqrt{3}\sqrt{\lambda^{-1} k^6 + 2}}{8\lambda^{-3/2} k^9 + 18 \lambda^{-1/2} k^3} \right) \right) - 2}{3\lambda k^{-6}} . \tag{6.36}$$

This quantity shows basically three different shapes:

$$\mathcal{P}(k) \to \begin{cases} \sim k^3 & : \quad k \ll k_{1/2} , \\ \text{const.} & : \quad k_{1/2} \ll k \ll 1/|\eta| , \\ \sim k^2 & : \quad 1/|\eta| \ll k . \end{cases} \tag{6.37}$$

Here $k_{1/2}$ is the argument value of $\mathcal{P}$ at which it has dropped to half of its asymptotic value. It is given by

$$k_{1/2} = \sqrt[6]{\frac{\lambda}{4}} . \tag{6.38}$$

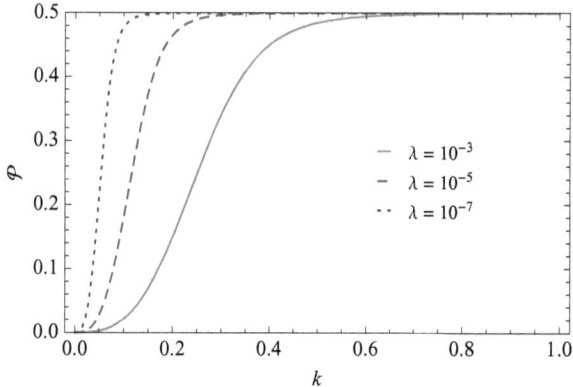

Figure 6.15: Power spectrum $\mathcal{P}(k)$ of a massless test field with quartic self-coupling as a function of comoving momentum k. Plotted are the results for various values of λ, where time has been fixed to $tH = 6$.

In the limit $\lambda \to 0$ it takes the constant value $1/2$, which is just the ordinary scale-free result obtained without any noise (in our normalisation). Hence, the result (6.36) is fully consistent with our previous work.

We observe in figure 6.15 that, depending on the coupling, the power spectrum is heavily suppressed on large scales. This damping becomes more pronounced as the coupling is increased. Thus, the region of broken scale invariance is pushed to smaller scales, which seems intuitive since the stronger the self-interaction the less (almost) free propagation is possible. Another way of seeing this is that the replica method applied in this work effectively resums arbitrary powers of the coupling λ, generating a self-energy σ, c.f. equation (5.25). Although dependent on space and time, this entity might be viewed as an effective mass term, which shifts the pole of the propagator G to non-zero values, resulting in a finite range of the inherent effective interaction.

6.5 Possible CMB effects

Recent investigations [HBB+96, SVP+03, CHS+06, CHS+08] show that there are virtually no large-scale correlations on the non-galactic cosmic microwave background. It is evident from figure 6.17 that above an angular scale of around $60°$ there is basically no correlation at all, if one excludes the largest angular scales ($\sim 160°$ - $180°$).

6.5 Possible CMB effects

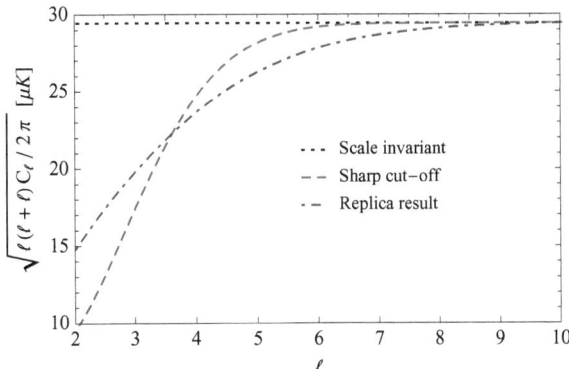

Figure 6.16: Comparison of different C_ℓ. The (black) dashed line represents the exact result from (6.41) for $\sqrt{\ell(\ell+1)C_\ell/2\pi}$ from a scale-invariant spectrum, which is constant. The (red) dashed curve displays the same quantity but evaluated with a sharp cut-off, and the (blue) dot-dashed line shows this for the power spectrum (6.36).

We recall that the two-point correlation function $\mathcal{C}(\theta)$ of the observed relative CMB temperature fluctuations $\Delta T := \delta T/T$ is defined and can be decomposed as

$$\mathcal{C}(\theta) := \left\langle \Delta T(\hat{e}_1)\,\Delta T(\hat{e}_2) \right\rangle_\theta = \frac{1}{4\pi} \sum_{\ell=0}^{\infty} (2\ell+1)\,C_\ell\,P_\ell(\cos\theta)\,, \tag{6.39}$$

where P_ℓ is the Legendre polynominal of degree ℓ and the angle brackets represent an average over all pairs of points on the sky (or at least the portion of the sky being analysed) that are separated by an angle θ. The coefficients $C_\ell \in \mathbb{R}$ are influenced by many effects, e.g. varying gravitational potential, reionisation, delayed recombination, polarisation, etc. We refer the interested reader to chapter 3 for details.

Because we expect drastic changes of $\mathcal{C}(\theta)$ due to the large-scale damping of $\mathcal{P}(k)$, and to have analytically tractable results, we only consider the leading ordinary Sachs-Wolfe effect [SW67]. Taking into account that the power spectrum (6.36) is actually a function of $\lambda^{1/6}\,k$, we find with the definition of the new coupling parameter

$$\lambda' := e^{-6\tilde{N}}\,[H\eta_0]^6\,\lambda \tag{6.40}$$

after a change of the integration variable from equation (3.36) for the multipole moments

$$C_\ell \approx A \int_0^\infty \frac{dk}{k}\,[j_\ell(k)]^2\,\mathcal{P}(k,\lambda')\,. \tag{6.41}$$

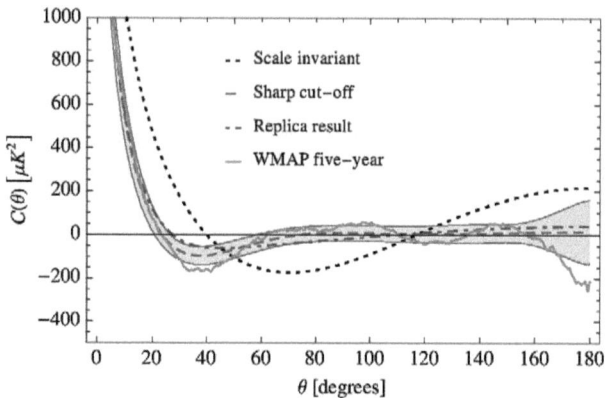

Figure 6.17: Angular temperature two-point correlation function $C(\theta)$, with the monopole and the dipole being subtracted. The (blue) dot-dashed line corresponds to the late-time power spectrum (6.36) and the (red) dashed curve represents the same quantity, but with a sharp cut at $k = k_{1/2}\eta_0 = 4.8$ to fit observations. Also plotted are the WMAP five-year data (ILC KQ75 mask) [KDN+09] (green, solid), the result for a scale-invariant power spectrum (black, dotted) and the one-sigma cosmic variance band around the sharp-cut curve (red-shaded).

Here j_ℓ is a spherical Bessel function of degree ℓ and η_0 is the conformal time of reception of CMB photons today. We define $A := 9/(25\pi)\Delta_{\mathcal{R}}^2 T_0^2 e^{-2\tau} \approx 1733.57\,\mu K^2$, where the values of the amplitude $\Delta_{\mathcal{R}}^2 \approx 2.45 \times 10^{-9}$ and the optical depth $\tau \approx 0.084$ have been taken from [KDN+09].

The factor $e^{-6\tilde{N}}$ in (6.40) arises from the conversion from comoving to physical momenta and depends on the specific model and parameters for the evolution of the Universe. If we assume as an extremely simple approximation a two-period model with an instantaneous transition from a de Sitter to a radiation-dominated stage, the factor $\tilde{N}$ is given by

$$\tilde{N} = N_\tau + \ln\left(\frac{T_{\text{reh}}}{T_0}\right), \tag{6.42}$$

where T_{reh} is the reheating temperature. The factor $N_\tau := (t_f - t_i)H$ is the total number of (inflationary) e-folds, where t_i and t_0 being the initial and final time of this interval, respectively.

In figure 6.16 we show a comparison of the quantity $\sqrt{\ell(\ell+1)C_\ell/2\pi}$ with C_ℓ subject to (6.41) for the cases of a scale-invariant power spectrum (black, dotted), the power spectrum from (6.36) with $k_{1/2} = 4.8\eta_0$ (blue, dot-dashed) and a scale-invariant power spectrum but with a sharp lower momentum cut-off at $k_{\text{out}} = 4.8\eta_0$ (red, dashed).

6.5 Possible CMB effects

The latter is suggested from the step-like form of $\mathcal{P}(k)$ obtained in this work (fig. 6.15). One sees in figure 6.16 that C_ℓ is damped especially for low ℓ if either a smooth or a step-like momentum cut-off is used. For $\ell \gtrsim 10$ all cases basically coincide.

Let us briefly mention that the sharp cut just mentioned may also correspond to some extend to certain models with non-simply connected space-times [ALS+04, ALS+05, ALS05a, ALS05b, AJL+08], where, e.g. in the torodial case, the length L of the fundamental domain gives — besides a discrete spectrum — the infra-red cut-off $k = 2\pi/L$.

Figure 6.17 shows our results for the two-point function $\mathcal{C}(\theta)$, which has been obtained from the full power spectrum (6.36) together with formulae (6.39) and (6.41), in which we sum up to $\ell = 200$, calculated with the NIntegrate function of MATHEMATICA 7. One clearly sees that it matches all significant features of the corresponding WMAP five-year curve. Remarkable: This is even better fulfilled for the sharp cut: The corresponding curve and the data basically lie on top of each other in the angular range $8° \lesssim \theta \lesssim 25°$. Moreover, almost the entire WMAP data is found inside the one-sigma cosmic variance band (red-shaded). Also the scale of around $70°$, above which $\mathcal{C}(\theta)$ remains close to zero, can quantitatively be very well reproduced. For the best-fit value of $k_{1/2}$ we find

$$k_{1/2}\eta_0 \approx 4.8 \, . \tag{6.43}$$

Inserting

$$\eta_0 \approx \frac{2}{\sqrt{\Omega_m}\, H_0} \, , \tag{6.44}$$

where $\Omega_m \approx 0.27$ [KDN+09] is the matter density of the Universe today, we obtain

$$k_{1/2} \approx 1.25 \, H_0 \approx 2.95 \times 10^{-4} \, \text{Mpc}^{-1} \, . \tag{6.45}$$

Now, equation (6.38) together with equation (6.40) provides a relation among this damping scale, the associated self-coupling and the total number of inflationary e-folds N_T. As $k_{1/2}$ and λ are fixed or at least constrained from observations, one may use this relation to determine N_T. We find for $\lambda = 10^{-13}$, $H_0 = 10^{-42}$ GeV, $H = 10^{17}$ GeV, $T_\text{reh} = 10^{15}$ GeV and $T_0 = 10^{-13}$ GeV the result

$$N_\text{T} = \ln\left(4\,(k_{1/2}\eta_0)(H\eta_0)\sqrt[6]{\frac{\lambda}{4}\frac{T_0}{T_\text{reh}}}\right) \approx 70.5 \, , \tag{6.46}$$

which is well consistent with the minium amount derived in section 1.3.2.

The conclusion that the observed low-multipole suppression is caused by a damping of the power spectrum might be weakened to some extend by the integrated Sachs-Wolfe effect [SW67], which — among others — has to be included in a full numerical study. In fact, Mortonson and Hu [MH09] recently commented on this issue. As follows from (6.45), their bound of $k_\text{cut} < 5.2 \times 10^{-4} \, \text{Mpc}^{-1}$ is in full agreement with our results.

Summary & Outlook

»*Das Ziel ist erreicht; doch auch aus dem Rückblick
schöpfe die Kraft zur Krönung des Werks.*«

according to WILHELM JORDAN

In this thesis we studied the large-scale behaviour of the power spectrum of the long-wavelength part of multi-component scalar test fields in curved space-time, using a stochastic description for the quantum modes as introduced by Starobinsky [Sta82]. We focused on the two important cases of a spatially-flat Friedmann geometry with an exponential and a power-law scale factor. The effective spectral index is calculated in the framework of replica field theory after Mézard and Parisi [MP91], which we recently introduced in a cosmological context [KS08, KS09]. Using a Gaussian variational approximation [Fey55], we derived an expression for the physical propagator G and thus for the power spectrum of long-wavelength fluctuations. These methods allow us to study the spatial behaviour of arbitrary long-wavelength two-point correlation functions.

A discussion on possible filter functions has been given with special focus on the aspect of the compactness of their support. For filter functions which deviate from the step function on an unbounded interval J, we find the phenomenon of dimensional reduction on super-horizon scales. It heavily amplifies the power spectrum in the infra-red. The time evolution of the long-wavelength field pushes the dimensionally-reduced region exponentially fast to unobservable scales. Taking the limit of vanishing width of the filter function, i.e. of a sharp separation of long- and short-wavelength modes, has a similar same effect. For filter functions with a bounded interval J, we show that the smooth separation might also lead to strong modifications of the power spectrum. However, this effect is limited to J and decreases exponentially fast, becoming negligible after a few e-folds.

Our findings provide further support for the self-consistency of the idea of inflation. Either regions of broken scale invariance with extraordinarily large fluctuations disappear faster than any causal patch of the universe expands (unbounded J), or large extra power is strongly damped by the time evolution (bounded J).

The scenario of eternal inflation [Vil83, LLM94], where huge fluctuations on large scales are responsible for the permanent creation of new Universes, is also qualitatively consistent with our results. However, when talking about the birth of new Universes, our set-up of a test field on a fixed background is surely wrong and a quantitative comparison is not possible.

The huge effect of quantum noise on large super-horizon scales (may they occur on a finite or infinite momentum range) does not permit us to speak about the spectrum of fluctuations in the usual, perturbative sense and clearly signals a breakdown of ordinary perturbation theory. This noise modifications further display the failure of the test-field assumption, since in the situation at hand it is no longer valid to neglect the back-reaction of the field on the geometry. In the case of bounded J it is possible to avoid the breakdown of perturbation theory and the test field assumption by an appropriate choice of the filter width.

We then applied our new methods to interacting scalar fields, where we focus on quartic self-coupling. There, it is found that the long-wavelength power spectrum acquires an additional damping on large scales, which can be understood from the effective mass generation due to the resummation of the stochastic quantum effects.

This damping is then applied to the description of the CMB data. Depending on the particular damping scale and by virtue of only the ordinary Sachs-Wolfe effect, one obtains a temperature auto-correlation function fairly close to the one observed by WMAP. As our derivation does not take into account the integrated Sachs-Wolfe effect, which might largely compensate the damping effect, a final statement can only be drawn from a full numerical study.

Another open task is to study more general space-times, such as those modelling inhomogeneities and to treat also the geometry stochastically. For a fully consistent treatment this is necessary because the Einstein field equations relate the energy densities, e.g. of the quantum field, to the curvature of the underlying space-time.

A further, rather obvious challenge would be to perform a full functional renormalisa-tion-group study [Fis85, Fis86, GL94, Fel00, LW02, LW04, LWC04, LW06], which is linked to replica field theory (c.f. e.g. [LMW08]). This approach allows for studying the (functional) beta function of the renormalised noise correlator $\tilde{R}$ and is thus — via a fixed point analysis — capable of determining the specific form of the physical correlator.

Appendix

Replica Symmetry Breaking

»*When the stars threw down their spears,*
And water'd heaven with their tears:
Did he smile his work to see?
Did he who made the Lamb make thee?

Tyger Tyger, burning bright,
In the forests of the night;
What immortal hand or eye,
Dare frame thy fearful symmetry?«

WILLIAM BLAKE

In the preceding chapters we saw the occurrence of the matrix σ which is used as a variational parameter. Even for very simple models like the Sherrington-Kirkpatrick model [SK75] it is impossible to treat the case where σ is arbitrary. So one is forced to make assumptions due to its structure. The easiest, the replica symmetric ansatz, has been introduced above. However, one can show that this ansatz produces in discrete models *a negative entropy* at zero temperature.

One thus looks in the treatment of disordered systems at low temperatures for a more fruitful ansatz. It consists no longer of a uniform replica matrix — that is why this cases is referred to as **replica symmetry breaking**. Due to the occurrence of ergodicity breaking at low temperatures T, which becomes stronger if T is lowered, we may find it, at least a bit, intuitive to give σ a hierarchical structure. By this we mean a matrix consisting of blocks inside blocks (c.f. the illustrative example on the subsequent page).

If we have k blocks, we talk about **k-step replica symmetry breaking**. Because the fragmentation of the phase space continues as the system is cooled down, one may take the limit $k \to \infty$ in which the *matrix* (σ_{ab}) becomes actually a *function* $\sigma(u)$. This is the case of so-called **full-step replica symmetry breaking**.

$$\sigma = \begin{pmatrix}
\sigma_c & \sigma_3 & \sigma_2 & \sigma_2 & \sigma_1 & \sigma_1 & \sigma_1 & \sigma_1 & \sigma_0 & \sigma_0 & \sigma_0 & \sigma_0 & \sigma_0 & \sigma_0 & \sigma_0 & \sigma_0 \\
\sigma_3 & \sigma_c & \sigma_2 & \sigma_2 & \sigma_1 & \sigma_1 & \sigma_1 & \sigma_1 & \sigma_0 & \sigma_0 & \sigma_0 & \sigma_0 & \sigma_0 & \sigma_0 & \sigma_0 & \sigma_0 \\
\sigma_2 & \sigma_2 & \sigma_c & \sigma_3 & \sigma_1 & \sigma_1 & \sigma_1 & \sigma_1 & \sigma_0 & \sigma_0 & \sigma_0 & \sigma_0 & \sigma_0 & \sigma_0 & \sigma_0 & \sigma_0 \\
\sigma_2 & \sigma_2 & \sigma_3 & \sigma_c & \sigma_1 & \sigma_1 & \sigma_1 & \sigma_1 & \sigma_0 & \sigma_0 & \sigma_0 & \sigma_0 & \sigma_0 & \sigma_0 & \sigma_0 & \sigma_0 \\
\sigma_1 & \sigma_1 & \sigma_1 & \sigma_1 & \sigma_c & \sigma_3 & \sigma_2 & \sigma_2 & \sigma_0 & \sigma_0 & \sigma_0 & \sigma_0 & \sigma_0 & \sigma_0 & \sigma_0 & \sigma_0 \\
\sigma_1 & \sigma_1 & \sigma_1 & \sigma_1 & \sigma_3 & \sigma_c & \sigma_2 & \sigma_2 & \sigma_0 & \sigma_0 & \sigma_0 & \sigma_0 & \sigma_0 & \sigma_0 & \sigma_0 & \sigma_0 \\
\sigma_1 & \sigma_1 & \sigma_1 & \sigma_1 & \sigma_2 & \sigma_2 & \sigma_c & \sigma_3 & \sigma_0 & \sigma_0 & \sigma_0 & \sigma_0 & \sigma_0 & \sigma_0 & \sigma_0 & \sigma_0 \\
\sigma_1 & \sigma_1 & \sigma_1 & \sigma_1 & \sigma_2 & \sigma_2 & \sigma_3 & \sigma_c & \sigma_0 & \sigma_0 & \sigma_0 & \sigma_0 & \sigma_0 & \sigma_0 & \sigma_0 & \sigma_0 \\
\sigma_0 & \sigma_0 & \sigma_0 & \sigma_0 & \sigma_0 & \sigma_0 & \sigma_0 & \sigma_0 & \sigma_c & \sigma_3 & \sigma_2 & \sigma_2 & \sigma_1 & \sigma_1 & \sigma_1 & \sigma_1 \\
\sigma_0 & \sigma_0 & \sigma_0 & \sigma_0 & \sigma_0 & \sigma_0 & \sigma_0 & \sigma_0 & \sigma_3 & \sigma_c & \sigma_2 & \sigma_2 & \sigma_1 & \sigma_1 & \sigma_1 & \sigma_1 \\
\sigma_0 & \sigma_0 & \sigma_0 & \sigma_0 & \sigma_0 & \sigma_0 & \sigma_0 & \sigma_0 & \sigma_2 & \sigma_2 & \sigma_c & \sigma_3 & \sigma_1 & \sigma_1 & \sigma_1 & \sigma_1 \\
\sigma_0 & \sigma_0 & \sigma_0 & \sigma_0 & \sigma_0 & \sigma_0 & \sigma_0 & \sigma_0 & \sigma_2 & \sigma_2 & \sigma_3 & \sigma_c & \sigma_1 & \sigma_1 & \sigma_1 & \sigma_1 \\
\sigma_0 & \sigma_0 & \sigma_0 & \sigma_0 & \sigma_0 & \sigma_0 & \sigma_0 & \sigma_0 & \sigma_1 & \sigma_1 & \sigma_1 & \sigma_1 & \sigma_c & \sigma_3 & \sigma_2 & \sigma_2 \\
\sigma_0 & \sigma_0 & \sigma_0 & \sigma_0 & \sigma_0 & \sigma_0 & \sigma_0 & \sigma_0 & \sigma_1 & \sigma_1 & \sigma_1 & \sigma_1 & \sigma_3 & \sigma_c & \sigma_2 & \sigma_2 \\
\sigma_0 & \sigma_0 & \sigma_0 & \sigma_0 & \sigma_0 & \sigma_0 & \sigma_0 & \sigma_0 & \sigma_1 & \sigma_1 & \sigma_1 & \sigma_1 & \sigma_2 & \sigma_2 & \sigma_c & \sigma_3 \\
\sigma_0 & \sigma_0 & \sigma_0 & \sigma_0 & \sigma_0 & \sigma_0 & \sigma_0 & \sigma_0 & \sigma_1 & \sigma_1 & \sigma_1 & \sigma_1 & \sigma_2 & \sigma_2 & \sigma_3 & \sigma_c
\end{pmatrix}. \quad (6.47)$$

Mathematics of Replica Symmetry Breaking

Introductions to replica symmetry breaking can be found e.g. in [MP91, Dot00, Wie02]. We mainly follow, and in part summarise, those references here with emphasis on basic aspects for the purpose of a mathematical specification.

To model the large amount of energy minima in disordered systems, one conveniently utilises so-called **Parisi matrices**

$$\mathbb{Q} = \tilde{q}\mathbb{1} + \sum_{j=0}^{k} \bar{\mathbb{1}}_j q_j , \qquad (6.48)$$

where $\bar{\mathbb{1}}_j := \mathbb{1}_j - \mathbb{1}_{j+1}$ and $\mathbb{1}_j := \mathbb{1}_{m_j}$ is the matrix that has blocks of size m_j on the main diagonal with every entry equal to 1 inside the blocks and 0 outside. The imposed restrictions are $\gcd(m_j, m_i) = \min(m_j, m_i)$ and $i < j$ implies $m_i > m_j$. Here we set $\mathbb{1}_0 := \mathbb{1}$ and $\mathbb{1}_{k+1} := 1$. Defining further the matrices

$$\mathbb{P}_j := \frac{1}{m_j}\mathbb{1}_j , \qquad \bar{\mathbb{P}}_j := \mathbb{P}_j - \mathbb{P}_{j-1} , \qquad \mathbb{P}_0 \equiv 0 , \qquad (6.49)$$

we see that the $\bar{\mathbb{P}}$s are projectors:

$$\bar{\mathbb{P}}_j \bar{\mathbb{P}}_i \equiv \delta_{ij} \bar{\mathbb{P}}_i \, , \qquad \sum_{j=1}^k \bar{\mathbb{P}}_j = \mathbb{1} \, . \qquad (6.50)$$

According to this, formula (6.48) can be written as

$$\mathbb{Q} = \tilde{q}\mathbb{1} + \sum_{j=0}^k q_j \left(m_j \, \mathbb{P}_j - m_{j+1} \, \mathbb{P}_{j+1} \right) . \qquad (6.51)$$

This matrix has the eigenvalues

$$\lambda = \tilde{q} + \sum_{j=0}^k q_j \left(m_j - m_{j+1} \right) \qquad \text{with multiplicity: } 1 \, , \qquad (6.52a)$$

$$\hat{q}_i = \tilde{q} + \sum_{j \geq i}^k q_j \left(m_j - m_{j+1} \right) - m_i \, q_{i-1} \qquad \text{with multiplicity: } n \frac{m_{i-1} - m_i}{m_{i-1} \, m_i} \, . \qquad (6.52b)$$

The last multiplicity is obtained by noting that

$$\dim \ker \left(\bar{\mathbb{P}}_i \right) = \dim \ker \left(\mathbb{P}_i - \mathbb{P}_{i-1} \right) = \frac{n}{m_i} - \frac{n}{m_{i-1}} = n \frac{m_{i-1} - m_i}{m_{i-1} \, m_i} \, . \qquad (6.53)$$

The continuum limit is achieved through

$$\sum_{j=0}^k q_j \left(m_j - m_{j+1} \right) = \sum_{(m_j)} q_{m_j} \left(m_j - m_{j+1} \right) \xrightarrow{m_j \to x} - \int_n^1 \mathrm{d}x \, q(x) \qquad (6.54)$$

and yields for (6.52a) and (6.52b)

$$\lambda = \tilde{q} - \langle q \rangle_n + n \, q(n) \qquad \text{with multiplicity: } 1 \, , \qquad (6.55a)$$

$$\hat{q}_n(x) = \tilde{q} - \langle q \rangle_n - [q]_n(x) \qquad \text{with multiplicity: } - n \frac{\mathrm{d}x}{x^2} \, . \qquad (6.55b)$$

The brackets are defined as follows

$$\langle q \rangle_n := \int_n^1 \mathrm{d}x \, q(x) + n \, q(n) \, , \qquad \langle \cdot \rangle_0 =: \langle \cdot \rangle \, , \qquad (6.56a)$$

$$[q]_n(x) := x \, q(x) - \int_n^x \mathrm{d}y \, q(y) - n \, q(n) \, , \qquad [\cdot]_0 =: [\cdot] \, . \qquad (6.56b)$$

Obviously we have

$$\frac{\mathrm{d}[q]_n(x)}{\mathrm{d}x} = x\, q'(x)\,, \tag{6.57a}$$

$$[q]_n(x) = 0 \quad\Leftrightarrow\quad q = \mathrm{const.}\,, \tag{6.57b}$$

$$[q]_n(n) = 0\,. \tag{6.57c}$$

Armed with this machinery one can easily invert a Parisi matrix:

$$\mathbb{1} = \mathbb{R}\,\mathbb{Q} = \mathbb{R}\,\mathbb{1}\,\mathbb{Q}\,\mathbb{1} = \sum_{i,j} \mathbb{R}\,\bar{\mathbb{P}}_i\,\mathbb{Q}\,\bar{\mathbb{P}}_j = \sum_{i,j}\hat{r}_i\,\bar{\mathbb{P}}_i\,\hat{q}_j\,\bar{\mathbb{P}}_j = \sum_i \hat{r}_i\,\hat{q}_i\,\bar{\mathbb{P}}_i \tag{6.58}$$

$$\Rightarrow \qquad 1 = \hat{r}_i\,\hat{q}_i \quad \forall\, i$$
$$\longrightarrow \quad 1 = \hat{q}(x)\,\hat{r}(x) = \big(\tilde{q} - \langle q\rangle - [q]\,(x)\big)\big(\tilde{r} - \langle r\rangle - [r]\,(x)\big)\,. \tag{6.59}$$

We can also calculate $\mathrm{tr}\,\ln(\mathbb{Q})$:

$$\frac{1}{n}\mathrm{tr}\,\ln(\mathbb{Q}) = \frac{1}{n}\ln\big(\tilde{q} - \langle q\rangle_n + n\,q(0)\big) - \underbrace{\int_n^1 \frac{\mathrm{d}y}{y^2}\ln\big(\tilde{q} - \langle q\rangle_n - [q]_n(x)\big)}_{\text{from multiplicity (6.55b)}}$$

$$= \ln\big(\tilde{q} - \langle q\rangle_n\big) + \frac{1}{n}\ln\left(1 + \frac{n\,q(0)}{\tilde{q} - \langle q\rangle_n}\right) - \int_n^1 \frac{\mathrm{d}y}{y^2}\ln\left(\frac{\tilde{q} - \langle q\rangle_n - [q]_n(x)}{\tilde{q} - \langle q\rangle_n}\right)$$

$$\xrightarrow{n\to 0} \ln\big(\tilde{q} - \langle q\rangle\big) + \frac{q(0)}{\tilde{q} - \langle q\rangle} - \int_0^1 \frac{\mathrm{d}y}{y^2}\ln\left(\frac{\tilde{q} - \langle q\rangle - [q]\,(x)}{\tilde{q} - \langle q\rangle}\right), \tag{6.60}$$

which is important for evaluation of determinants arising form Gaussian path integrals. (6.60) is exactly the result found by Mézard and Parisi [MP91]. After some work one finds for the inverse Parisi matrix $\mathbb{P} = \big(\tilde{p}, p(x)\big) = \big(\tilde{q}, q(x)\big)^{-1} = \mathbb{Q}^{-1}$ in the continuum limit

$$p(x) = -\frac{1}{\tilde{q} - \langle q\rangle}\left(\frac{[q]\,(x)}{x\,[\tilde{q} - \langle q\rangle - [q]\,(x)]} + \frac{q(0)}{\tilde{q} - \langle q\rangle} + \int_0^x \frac{\mathrm{d}y}{y^2}\frac{[q]\,(y)}{\tilde{q} - \langle q\rangle - [q]\,(y)}\right), \tag{6.61a}$$

$$\tilde{p} = \frac{1}{\tilde{q} - \langle q\rangle}\left(1 - \frac{q(0)}{\tilde{q} - \langle q\rangle} - \int_0^1 \frac{\mathrm{d}y}{y^2}\frac{[q]\,(y)}{\tilde{q} - \langle q\rangle - [q]\,(y)}\right), \tag{6.61b}$$

$$\Rightarrow \quad \tilde{p} - p(x) = \frac{1}{x\,[\tilde{q} - \langle q\rangle - [q]\,(x)]} - \int_x^1 \frac{\mathrm{d}y}{y^2}\frac{1}{\tilde{q} - \langle q\rangle - [q]\,(y)} \tag{6.61c}$$

and for the product of two Parisi matrices, $\mathbb{R} = \mathbb{P}\mathbb{Q}$, (c.f. [Dot00])

$$\tilde{r}_n = \tilde{p}\tilde{q} - \int_n^1 dx\, p(x)\,q(x)\,, \tag{6.62a}$$

$$r_n(x) = -n\,a(x)\,b(x) + \left(\tilde{p} - \langle p\rangle_n + n\,p(n)\right) q(x) + \left(\tilde{q} - \langle q\rangle_n + n\,q(n)\right) p(x)$$
$$- \int_n^x dy\, \bigl(p(y) - p(x)\bigr)\bigl(q(y) - q(x)\bigr)\,. \tag{6.62b}$$

Thus for $n \to 0$ the product of three such matrices, $\mathbb{G} = \mathbb{M}\mathbb{P}\mathbb{Q}$, with $\tilde{m} = \tilde{p} = \tilde{q} \stackrel{!}{=} 0$ is

$$\tilde{g} = \int_0^1 dx \left(x\, m(x)\, p(x)\, q(x) + m(x) \int_x^1 dy\, p(y)\, q(y) \right.$$
$$\left. + q(x) \int_x^1 dy\, p(y)\, m(y) + p(x) \int_x^1 dy\, m(y)\, q(y) \right). \tag{6.63}$$

Using this one can obtain information of the structure of the underlying space. Therefore we introduce the **Sherrington-Kirkpatrick model** [SK75], which is defined via

$$\mathcal{H} := -\sum_{i<j}^N J_{ij}\sigma_i\sigma_j\,, \qquad \mathcal{P}[J_{ij}] := \prod_{i<j} \sqrt{\frac{N}{2\pi}} \exp\left(-\frac{N}{2} J_{ij}\right), \tag{6.64}$$

with the **spin variables** $\sigma_j \in\,]-1,+1[$. The factor N in the exponential in front of the J is essential for the free energy to be extensive. A bit of work yields [Dot00]

$$\overline{Z^n} \propto \prod_{a<b} \sum_{\{\sigma\}} \int_{-\infty}^{+\infty} dQ_{ab}^n\, \exp\left(\frac{\beta^2 n(N-n)}{4} + \beta^2 \sum_{\substack{a<b \\ j\in N}} Q_{ab}^{(n)} \sigma_j^a \sigma_j^b - \frac{\beta^2 N}{2} \sum_{a<b} \left(Q_{ab}^{(n)}\right)^2 \right). \tag{6.65}$$

After the variation

$$\frac{\delta \overline{Z^n}}{\delta Q_{ab}^{(n)}} = 0 \tag{6.66}$$

we have the important equation

$$Q_{ab}^{(n)} = \frac{1}{N} \sum_{j=1}^N \overline{\langle \sigma_j^a \sigma_j^b \rangle}\,. \tag{6.67}$$

For spin variables σ_j define the **pure states** α with weights ω_α by the property

$$\langle \sigma_j \sigma_k \rangle_\alpha := \langle \sigma_j \rangle_\alpha \langle \sigma_k \rangle_\alpha \tag{6.68}$$

for which

$$\langle \cdot \rangle \equiv \sum_\alpha \omega_\alpha \langle \cdot \rangle_\alpha . \tag{6.69}$$

Furthermore we define the **overlap** $q_{\alpha\beta}$ via

$$q_{\alpha\beta} := \frac{1}{N} \sum_{j=1}^{N} \langle \sigma_i \rangle_\alpha \langle \sigma_i \rangle_\beta \tag{6.70}$$

and, to describe the statistics of the overlap, the **probability distribution function**

$$P(q) := \overline{\sum_{\alpha\beta} \omega_\alpha \omega_\beta \delta(q - q_{\alpha\beta})} . \tag{6.71}$$

As mentioned previously, it is the probability of finding two replicas having their overlap equal to q. It is precisely this function which one can consider as an order parameter. Victor Dotsenko said: »*The fact that it is a function is actually a manifestation of the crucial phenomenon that for the description of the spin glass phase we need an infinite number of order parameters.*« [Dot00]. Defining for all $k \in \mathbb{N}$ the quantities

$$q^{(k)} := \frac{1}{N^k} \sum_{i_1 \ldots i_k = 1}^{N} \overline{\left\langle \prod_{j=1}^{k} \sigma_{ij} \right\rangle^2} \tag{6.72a}$$

we see, using (6.68) and (6.69), that

$$q^{(k)} := \int dq \, P(q) \, q^k , \tag{6.72b}$$

which means, that these parameters are the **moments of the probability density** P and that »*the function $P(q)$ originally defined to describe the statistics of (somewhat abstract) pure states, can be calculated (at least theoretically) from the multipoint correlation functions in the Gibbs states.*« [Dot00]

We note that the probability that three arbitrary pure states having mutual overlaps $q_{\alpha\beta}$, $q_{\alpha\gamma}$ and $q_{\beta\gamma}$ equal to q_1, q_2 and q_3 is

$$P(q_1, q_2, q_3) = \overline{\sum_{\alpha\beta\gamma} \omega_\alpha \omega_\beta \omega_\gamma \delta(q_1 - q_{\alpha\beta}) \delta(q_2 - q_{\alpha\gamma}) \delta(q_3 - q_{\beta\gamma})} \tag{6.73a}$$

and can be written as (c.f. [Dot00])

$$P(q_1, q_2, q_3) = \lim_{n \to 0} \underbrace{\frac{1}{n(n-1)(n-2)}}_{= \# \text{ of terms in the sum}} \sum_{\substack{a,b,c \\ a \neq b \neq c \neq a}} \delta(q_1 - Q_{ab}^n) \delta(q_2 - Q_{ac}^n) \delta(q_3 - Q_{bc}^n) . \tag{6.73b}$$

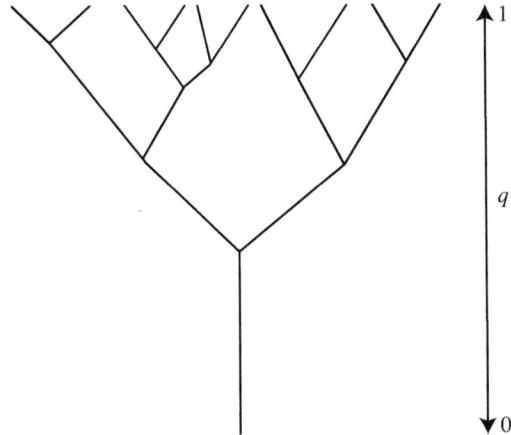

Figure 6.18: An example of an ultrametrical tree.

With formula (6.63) we obtain from $\frac{d}{dq}x(q) = P(q)$ the result

$$P(q_1, q_2, q_3) = \frac{1}{2} P(q_1) x(q_1) \delta(q_1 - q_2) \delta(q_1 - q_3) + \frac{1}{2} P(q_1) P(q_2) \theta(q_1 - q_2) \delta(q_2 - q_3) \\ + \frac{1}{2} P(q_2) P(q_3) \theta(q_2 - q_3) \delta(q_1 - q_3) + \frac{1}{2} P(q_3) P(q_1) \theta(q_1 - q_3) \delta(q_1 - q_2). \quad (6.74)$$

This shows the characteristic topological property of **ultrametricity**, which is that of the endpoints of a hierarchical tree (c.f. [RTV86, Dot00] for more detailed discussions).

A Replica Symmetry-Breaking Example

Except from the previous section, we studied cases of replica symmetry. Since for more sophisticated random potentials this is only an *assumption* which might induce instabilities under certain conditions (see e.g. [MP91]), we will now discuss the scenario where replica symmetry is broken. It implies that the matrix σ is no longer uniform in replica space. The case we will treat here is continuous replica symmetry breaking. Already involved enough we will restrict ourselves to the short-range case throughout this section.

We stress that our motivation is of purely theoretical nature — we would like to understand what happens in the case of continuously broken replica symmetry, probably arising from some more fundamental theory. Let us work in d space-time dimensions and focus on correlators $\widehat{R}$ of the form $\widehat{R}(z) \propto z^{-\gamma}$, with $\gamma \in \mathbb{R}$.

According to section 6.5 we can, in the limit $n \to 0$, parametrise a Parisi matrix by a function $\sigma : [0, 1] \to \mathbb{R}$ (the off-diagonal elements), and a constant σ_c (the part proportional to $\mathbb{1}$). In analogy we have for the propagators

$$G_{aa}(t,k) \mapsto \tilde{g}(t,k), \quad G_{ab}(t,k) \mapsto g(t,k,u). \tag{6.75}$$

Then, using (5.18a), the variational equations (5.30a) and (5.30b) give in the limit $n \to 0$

$$\sigma_c(t) = -\widehat{R}'\left(\int_k \tilde{g}(t,k)\right), \tag{6.76a}$$

$$\sigma(t,u) = \widehat{R}'\left(\int_k g(t,k,u)\right). \tag{6.76b}$$

Using (5.18b) yields for (5.32a)

$$\sigma(t,u) = -2\widehat{R}'\left(2\int_k \left(\tilde{g}(t,k) - g(t,k,u)\right)\right). \tag{6.77}$$

Let us define $C(t,k) := G_0^{-1}(t,k) + M(t)$ with $M(t) := \sigma_c(t) + \langle \sigma \rangle(t)$. Of course, $M(t) = 0$ in the case of difference correlation if $U \equiv 0$. With (see (6.61b))

$$\tilde{g}(t,k) = \frac{1}{C(t,k)}\left(1 + \frac{\sigma(t,0)}{C(t,k)} + \int_0^1 \frac{dy}{y^2} \frac{[\sigma](t,y)}{C(t,k) + [\sigma](t,y)}\right), \tag{6.78a}$$

$$g(t,k,u) = \frac{1}{C(t,k)}\left(\frac{[\sigma](t,u)}{u[C(t,k) + [\sigma](t,u)]} + \frac{\sigma(t,0)}{c(t)\,k^{2\alpha} + M(t)} + \int_0^u \frac{dy}{y^2} \frac{[\sigma](t,y)}{C(t,k) + [\sigma](t,y)}\right) \tag{6.78b}$$

we have for (6.76a) and (6.76b)

$$\sigma_c(t) = -\widehat{R}'\left(\int_k \frac{1}{C(t,k)}\left(1 + \frac{\sigma(t,0)}{C(t,k)} + \int_0^1 \frac{dy}{y^2} \frac{[\sigma](t,y)}{C(t,k) + [\sigma](t,y)}\right)\right), \tag{6.79a}$$

$$\sigma(t,u) = \widehat{R}'\left(\int_k \frac{1}{C(t,k)}\left(\frac{[\sigma](t,u)}{u[C(t,k) + [\sigma](t,u)]} + \frac{\sigma(t,0)}{C(t,k)} + \int_0^u \frac{dy}{y^2} \frac{[\sigma](t,y)}{C(t,k) + [\sigma](t,y)}\right)\right). \tag{6.79b}$$

Equation (6.77) becomes (c.f. (6.61c))

$$g(t,k,u) = \frac{1}{u\big[C(t,k) + [\sigma](t,u)\big]} - \int_u^1 \frac{dy}{y^2} \frac{1}{C(t,k) + [\sigma](t,y)}, \qquad (6.80)$$

and thus

$$\sigma(t,u) = -2\widehat{R}'\left(2\int_k \left(\frac{1}{u\big[C(t,k)+[\sigma](t,u)\big]} - \int_u^1 \frac{dy}{y^2}\frac{1}{C(t,k)+[\sigma](t,y)}\right)\right). \qquad (6.81)$$

Differentiating (6.81) once with respect to u yields

$$\sigma'(u) = \widehat{R}''(\ldots) \int_k \frac{1}{\big[C(t,k)+[\sigma](t,u)\big]^2} \sigma'(u), \qquad (6.82)$$

so we have one replica-symmetric solution and, if $\sigma' \neq 0$, another one given by

$$\frac{1}{\widehat{R}''(\ldots)} = \int_k \frac{1}{\big[C(t,k)+[\sigma](t,u)\big]^2}. \qquad (6.83)$$

Inverting (6.83), inserting this into (6.76b) and assuming the existence of R''^{-1} yields in analogy to [LW03] some very general formulae

$$\sigma(u,t) = \widehat{R}'\left(\big[\widehat{R}''\big]^{-1}\left(\frac{1}{\int_k \frac{1}{[C(t,k)+[\sigma](t,u)]^2}}\right)\right), \qquad (6.84a)$$

$$u(t) = \frac{\left[\int_k \frac{1}{[C(t,k)+[\sigma](t,u)]^2}\right]^3}{\int_k \frac{1}{[C(t,k)+[\sigma](t,u)]^3}} \widehat{R}'''\left(\big[\widehat{R}''\big]^{-1}\left(\frac{1}{\int_k \frac{1}{[C(t,k)+[\sigma](t,u)]^2}}\right)\right). \qquad (6.84b)$$

Once the $\widehat{R}'$ is known, one can determine the complete function $[\sigma](t,u)$. It is clear from (6.83) that random anisotropy does not provide continuous replica symmetry breaking, neither for product nor for difference correlation. Now let $d < 1 + 4\alpha$ and assume $G_0^{-1}(t,k) = c(t)\,k^{2\alpha}$ and

$$\widehat{R}(z) \propto z^{-\gamma}. \qquad (6.85)$$

Thus, we see that equation (6.83) implies

$$(\ldots)^{\gamma+2} \propto \big[M(t) + [\sigma](t,u)\big]^{\frac{d-1-4\alpha}{2\alpha}}, \qquad (6.86)$$

where we used

$$\int \mathrm{d}^{d-1}k \, \frac{1}{[k^{2\alpha}+\rho]^\zeta} \propto \rho^{\frac{d-1-2\alpha\zeta}{2\alpha}} . \tag{6.87}$$

Differentiating (6.86) again with respect to u we find after some lines of calculation

$$M(t) + [\sigma](t,u) \propto u^{\frac{2\alpha(\gamma+2)}{d+(d-2\alpha-1)\gamma-1}} . \tag{6.88}$$

This and (6.78a) imply for $k \to 0$

$$\tilde{g}(t,k) \sim k^{-2\alpha\left(1+\frac{d+(d-2\alpha-1)\gamma-1}{2\alpha(\gamma+2)}\right)} = k^{1-d-\frac{4\alpha-d+1}{\gamma+2}} \equiv k^{-(d-1)} k^{n_s-1} \tag{6.89}$$

and hence for the spectral index

$$n_s = \frac{d-4\alpha+\gamma+1}{\gamma+2} . \tag{6.90}$$

Thus one can only achieve scale invariance only for $\gamma = -1 - \frac{4\alpha}{d-1}$ (being negative), or in the limit $\gamma \to \infty$. However, scale invariance is observationally tightly constrained: Experimental bounds, coming from five-year data of WMAP and other surveys suggest $n_\varphi \simeq 1$ [KDN+09]. This means that, assuming scale invariance is broken in a narrow range only, γ will be very large. Let us in general write $\Delta_l \leq n_\varphi \leq \Delta_u$ and ask what this implies for γ. For general α we find

$$\gamma \geq \min\left\{\frac{1-d+4\alpha}{\Delta_l} - 2, \frac{1-d+4\alpha}{\Delta_u} - 2\right\}, \tag{6.91a}$$

$$\gamma \leq \max\left\{\frac{1-d+4\alpha}{\Delta_l} - 2, \frac{1-d+4\alpha}{\Delta_u} - 2\right\}. \tag{6.91b}$$

This means in particular that measurements might directly constrain the noise correlator.

It is worth noting that for $\gamma = -1$ the replica symmetric results are recovered (as it should be for the random-field case). It is also interesting to compare our results to those of Mézard and Parisi: For $\gamma \to \gamma - 1$, $d \to d+1$ and $\alpha = 1$ the results reduce exactly to those found in [MP91].

The above derivations rely heavily on the specific form of the correlator R. Furthermore, we saw that power-law correlation does not allow for scale invariance. Nevertheless, the limit $\gamma \to \infty$ of formula (6.90) reproduces $n_\varphi = 1$, suggesting to study an exponential scenario, i.e.

$$\widehat{R}(x) \propto e^{-\chi x} \tag{6.92}$$

with $\chi \in \mathbb{R}$. Performing an analogous derivation as with the correlator (6.85) yields indeed a scale-invariant spectrum, i.e. $n_s = 1$, as anticipated.

Bibliography

[AIM76]   A. Aharony, Y. Imry and S.-k. Ma,
Lowering of Dimensionality in Phase Transitions ...
Phys. Rev. Lett. **37**, 1364 (1976)

[AJL+08]   R. Aurich, H. S. Janzer, S. Lustig and F. Steiner,
Do we Live in a "Small Universe"?
Class. Quant. Grav. **25**, 125006 (2008)

[ALS+04]   R. Aurich, S. Lustig, F. Steiner and H. Then,
Hyperbolic Universes with a Horned Topology and the CMB ...
Class. Quant. Grav. **21**, 4901 (2004)

[ALS+05]   R. Aurich, S. Lustig, F. Steiner and H. Then,
Can one Hear the Shape of the Universe?
Phys. Rev. Lett. **94**, 21301 (2005)

[ALS05a]   R. Aurich, S. Lustig and F. Steiner,
CMB Anisotropy of the Poincaré Dodecahedron,
Class. Quant. Grav. **22**, 2061 (2005)

[ALS05b]   R. Aurich, S. Lustig and F. Steiner,
CMB Anisotropy of Spherical Spaces,
Class. Quant. Grav. **22**, 3443 (2005)

[Ami84]   D. J. Amit,
Field Theory, the Renormalization Group and Critical Phenomena,
Revised Second Edition, World Scientific (1984)

[And76]   J. L. Anderson,
The Equations of Radiative Hydrodynamics,
Gen. Rel. Grav. **7**, 53 (1976)

[AP80]   A. Aharony and E. Pytte,
Infinite Susceptibility Phase in Random Uniaxial Anisotropy ...
Phys. Rev. Lett. **45**, 1583 (1980)

[AW89]   M. Aizeman and J. Wehr,
Rounding of First-Order Phase Transitions in Systems ...
Phys. Rev. Lett. **62**, 2503 (1989)

[BCD87]   B. Barbara, M. Coauch and B. Dieny,
          Low-Temperature Dynamics and Lower Critical Dimensionality ...
          Europhys. Lett. **3**, 1129 (1987)

[BCS95]   T. Bellini, N. A. Clark and D. W. Schaefer,
          Dynamic Light Scattering Study of Nematic and Smectic: ...
          Phys. Rev. Lett. **74**, 2740 (1995)

[Bel98]   D. P. Belanger,
          Experiments on the Random Field Ising Model,
          in *Spin Glasses and Random Fields*, World Scientific (1998)

[Bel00]   M. Bellini,
          Gauge-Invariant Fluctuations of the Metric in Stochastic Inflation,
          Phys. Rev. D **61**, 107301 (2000)

[Bel01]   M. Bellini,
          Coarse-Grained Field Wave Function in Stochastic Inflation,
          Nucl. Phys. B **604**, 441 (2001)

[Ber85]   A. Beretti,
          Some Properties of Random Ising Models,
          J. Stat. Phys. **38**, 483 (1985)

[BF93]   L. Balents and D. S. Fisher,
          Large-N Expansion of $(4-\varepsilon)$-Dimensional Oriented Manifolds ...
          Phys. Rev. B **48**, 5949 (1993)

[BFG+94] G. Blatter, M. V. Feigel'man, V. B. Geshkenbein, et al.
          Vortices in High-Temperature Superconductors,
          Rev. Mod. Phys. **66**, 1125 (1994)

[BMM+88] C. M. Bender, K. A. Milton, M. Moshe, et al.
          Novel Perturbative Scheme in Quantum Field Theory,
          Phys. Rev. **D**, 1472 (1988)

[CBM+93] N. A. Clark, T. Bellini, R. M. Malzbender, et al.
          X-Ray Scattering Study of Smectic Ordering in a Silica Aerogel,
          Phys. Rev. Lett. **71**, 3505 (1993)

[CC95]   T. Castellani and A. Cavagna,
          Spin-Glass Theory for Pedestrians,
          J. Stat. Mech., 5012 (2005)

[CHS+06]	C. J. Copi, D. Huterer, D. J. Schwarz and G. D. Starkman, *On the Large-Angle Anomalies of the Microwave Sky*, Mon. Not. Roy. Astron. Soc. **367**, 79 (2006)
[CHS+08]	C. J. Copi, D. Huterer, D. J. Schwarz and G. D. Starkman, *No Large-Angle Correlations on the Non-Galactic Microwave Sky*, astro-ph/**0808.3767** (2008)
[CLG96]	D. Carpentier, P. Le Doussal and T. Giamarchi, *Stability of the Bragg Glass Phase in a Layered Geometry*, Europhys. Lett. **35**, 379 (1996)
[Dav81]	F. David, *Cancellations of Infrared Divergences in the Two-Dimensional . . .* Commun. Math. Phys. **81**, 149 (1981)
[deG84]	P. G. de Gennes, *Liquid-Liquid Demixing Inside a Rigid Network. Qualitative . . .* J. Phys. Chem. **88**, 6469 (1984)
[Dot00]	V. Dotsenko, *Introduction to Statistical Mechanics of Disordered Spin Systems*, Camebridge University Press (2000)
[dT78]	J. R. L. de Almeida and D. J. Thouless, *Stability of the Sherrington-Kirkpatrick Solution . . .* J. Phys. A **11**, 983 (1978)
[DW87]	S. B. Dierker and P. Wiltzius, *Random-Field Transition of a Binary Liquid in a Porous Medium*, Phys. Rev. Lett. **58**, 1865 (1987)
[Emi99]	T. Emig, *Comment on "Order-Disorder Transition in an External Field . . ."* Phys. Rev. Lett. **82**, 3380 (1999)
[EN98]	T. Emig and T. Nattermann, *Roughening Transition of Interfaces in Disordered Systems*, Phys. Rev. Lett. **81**, 1469 (1998)
[ES04]	K. Enqvist and M. S. Sloth, *Possible Connection between the Location of the Cut-Off . . .* Phys. Rev. Lett. **93**, 221302 (2004)

[FA79]   S. Fishman and A. Aharony,
         Random Field Effects in Disordered Anisotropic Antiferromagnets,
         J. Phys. C **12**, 729 (1979)

[Fel97] D. E. Feldman,
         Weak Disorder in the Two-Dimensional XY Dipole Ferromagnet,
         Phys. Rev. B **56**, 3167 (1997)

[Fel00] D. E. Feldman,
         Lattice Dynamics and Reduced Thermal Conductivity ...
         Phys. Rev. B **61**, 9209 (2000)

[Fey55] R. P. Feynman,
         Slow Electrons in a Polar Crystal,
         Phys. Rev. **97**, 660 (1955)

[Fis85] D. S. Fisher,
         Random Fields, Random Anisotropies, Nonlinear σ Models, ...
         Phys. Rev. B **31**, 7233 (1985)

[Fis86] D. S. Fisher,
         Interface Fluctuations in Disordered Systems: $5 - \varepsilon$-Expansion ...
         Phys. Rev. Lett. **56**, 1964 (1986)

[FK07]   A. A. Fedorenko and F. Kühnel,
         Long-Range Correlated Random Field and Random Anisotropy ...
         Phys. Rev. B **75**, 174206 (2007)

[FLW06] A. A. Fedorenko, P. Le Doussal and K. J. Wiese,
         Statics and Dynamics of Elastic Manifolds in Media ...
         Phys. Rev. E **74**, 61109 (2006)

[Fri79] D. Friedan,
         Geometric Models for Critical Systems in $2 + \varepsilon$ Dimensions,
         talk given at the Nuffield Workshop on Quantum Gravity (1979)

[FT86]   M. V. Feigel'man and M. V. Tsodyks,
         Amorphous Magnets with Strong Random Anisotropy,
         Zh. Eksp. Teor. Fiz. **91**, 955 (1986)

[FT97]   S. V. Fridrikh and E. M. Terentjev,
         Order-Disorder Transition in an External Field ...
         Phys. Rev. Lett. **79**, 4661 (1997)

[Fro84]	J. Fröhlich, *Mathematical Aspects of the Physics of Disordered Systems*, in *The 1984 Les Houches Summer School*, New York, Plenum (1984)
[GAS96]	A. M. Gutin, V. I. Abkevich, E. I. Shakhnovich, *Cooperativity of Protein Folding and the Random-Field Ising Model*, cond-mat/**9606136** (1996)
[GH96]	M. J. P. Gingras and D. A. Huse, *Topological Defects in the Random-Field XY Model ...* Phys. Rev. B **53**, 15193 (1996)
[Gia03]	T. Giamarchi, *Statistical Field Theory*, lecture notes, University of Geneva (2003)
[Gin81]	S. L. Ginzburg, *Spin Glasses with Random Anisotropic Exchange*, Zh. Eksp. Teor. Fiz. **80**, 244 (1981)
[GIO96]	T. Garel, G. Iori and H. Orland, *Variational Study of the Random-Field XY Model*, Phys. Rev. B **53**, 2941 (1996)
[GL94]	T. Giamarchi and P. Le Doussal, *Elastic Theory of Pinned Flux Lattices*, Phys. Rev. Lett. **72**, 1530 (1994)
[GLM87]	A. S. Goncharov, A. D. Linde and V. F. Mukhanov, *The Global Structure of the Inflationary Universe*, Int. J. Mod. Phys. A **2**, 561 (1987)
[GMP+87]	J. T. Graham, M. Maliepaard, J. H. Page, et al. *Random-Field Effects on Ising Jahn-Teller Phase Transitions*, Phys. Rev. B **35**, 2098 (1987)
[Gol61]	J. Goldstone, *Field theories with »Superconductor« solutions*, Nuovo Cim. **19**, 154 (1961)
[Gol83]	Y. Y. Goldschmidt, *Magnets with Random Uniaxial Anisotropy: Thermodynamic ...* Nucl. Phys. B **225**, 123 (1983)

[Gut80]	A. H. Guth, *Inflationary Universe: A Possible Solution to the Horizon ...* Phys. Rev. D **23**, 347 (1981)
[HBB+96]	G. Hinshaw, A. J. Banday, C. L. Bennett, K. M. Gorski, ... *2-Point Correlations in the* COBE DMR *4-Year Anisotropy Maps,* ApJ. **464**, 25 (1996)
[HG97]	H. Haga and C. W. Garland, *Effect of Silica Aerosil Particles on Liquid-Crystal ...* Liq. Cryst. **22**, 275 (1997)
[HPZ73]	R. Harris, M. Plischke and M. J. Zuckermann, *New Model for Amorphous Magnetism,* Phys. Rev. Lett. **31**, 160 (1973)
[HY05]	T. Hattori and K. Yamamoto, *Non-Gaussianity in Multi-Field Stochastic Inflation ...* JCAP **507**, 5 (2005)
[IM75]	Y. Imry and S. K. Ma, *Random-Field Instability of the Ordered State ...* Phys. Rev. Lett. **35**, 1399 (1975)
[IS99]	A. V. Izyumov and K. V. Samokhin, *Field Theory of Self-Avoiding Walks in Random Media,* J. Phys. A **32**, 7842 (1999)
[Kan89]	H. E. Kandrup, *Stochastic Inflation as a Time-Dependent Random Walk,* Phys. Rev. B **39**, 2245 (1989)
[KDN+09]	E. Komatsu, J. Dunkley, M. R. Nolta, et al. *Five-Year Wilkinson Microwave Anisotropy Probe Observations: ...* ApJ. **180**, 330 (2009)
[KJK84]	A. Khurana, A. Jagannathan and J. M. Kosterlitz, *A Random Anisotropy Model: $1/N$-Expansion for Gaussian ...* Nucl. Phys. B **240**, 1 (1984)
[KLP84]	A. Klein, L. J. Landau and J. F. Perez, *Supersymmetry and the Parisi-Sourlas Dimensional Reduction: ...* Commun. Math. Phys. **94**, 459 (1984)

[KMT83]	M. Kardar, B. McClain and C. Taylor, *Dimensional Reduction with Correlated Random Fields. ...* Phys. Rev. B **27**, 5875 (1983)
[KNH97]	J. Kierfeld, T. Nattermann and T. Hwa, *Topological Order in the Vortex-Glass Phase ...* Phys. Rev. B **55**, 626 (1997)
[Kor93]	S. E. Korshunov, *Replica Symmetry Breaking in Vortex Glasses*, Phys. Rev. B **48**, 3969 (1993)
[KPP+97]	K. Matsumoto, J. V. Porto, L. Pollak, et al. *Quantum Phase Transition of ^{3}He in Aerogel at a Nonzero Pressure*, Phys. Rev. Lett. **79**, 253 (1997)
[KS01]	E. Komatsu and D. N. Spergel, *Acoustic Signatures in the Primary Microwave Background ...* Phys. Rev. D **63**, 63002 (2001)
[KS08]	F. Kühnel and D. J. Schwarz, *Stochastic Inflation and Dimensional Reduction*, Phys. Rev. D **78**, 103501 (2008)
[KS09]	F. Kühnel and D. J. Schwarz, *Stochastic Inflation and Replica Field Theory*, Phys. Rev. D **79**, 44009 (2009)
[LAMDA]	Web page of NASA's data center for CMB research, five-year WMAP data products: http://lambda.gsfc.nasa.gov (July 2009)
[Lar70]	A. I. Larkin, *Effect of Inhomogeneities on the Structure of the Mixed State ...* Zh. Eksp. Teor. Fiz. **58**, 1466 (1970)
[LeD04]	P. Le Doussal, *Functional Renormalization*, talk given at the Windsor Summer School (2004)
[LL03]	A. R. Liddle and S. M. Leach, *How Long before the End of Inflation were Observable ...* Phys. Rev. D **68**, 103503 (2003)

[LLM+02]   S. M. Leach, A. R. Liddle, J. Martin and D. J. Schwarz,
           Cosmological Parameter Estimation and the Inflationary Cosmology,
           Phys. Rev. D **66**, 23515 (2002)

[LLM94]   A. Linde, D. Linde and A. Mezhlumian,
           From the Big Bang Theory to the Theory of a Stationary Universe,
           Phys. Rev. D **49**, 1783 (1994)

[LMM+04]   M. Liguori, S. Matarrese, M. Musso and A. Riotto,
           Stochastic Inflation and the Lower Multipoles ...
           JCAP **408**, 11 (2004)

[LMW08]   P. Le Doussal, M. Müller, and K. J. Wiese,
           Cusps and Shocks in the Renormalized Potential of Glassy ...
           Phys. Rev. B **77**, 64203 (2008)

[LW02]   P. Le Doussal and K. J. Wiese,
           Functional Renormalization Group at Large N ...
           Phys. Rev. Lett. **89**, 125702 (2002)

[LW03]   P. Le Doussal and K. J. Wiese,
           Functional Renormalization Group at Large N ...
           Phys. Rev. B **68**, 174202 (2003)

[LW04]   P. Le Doussal and K. J. Wiese,
           Derivation of the Functional Renormalization Group β-Function ...
           Nucl. Phys. B **701**, 409 (2004)

[LWC04]   P. Le Doussal, K. J. Wiese and P. Chauve,
           Functional Renormalization Group and the Field Theory ...
           Phys. Rev. E **69**, 26112 (2004)

[LW06]   P. Le Doussal and K. J. Wiese,
           Random-Field Spin Models beyond 1 Loop: A Mechanism ...
           Phys. Rev. Lett. **96**, 197202 (2006)

[LW07]   M. Li and Y. Wang,
           A Stochastic Measure for Eternal Inflation,
           JCAP **708**, 7 (2007)

[Muk05]   V. Mukhanov
           Physical Foundations of Cosmology,
           Cambridge University Press (2005)

Bibliography

[MB06]   J. E. Madriz Aguilar and M. Bellini,
Stochastic Gravitoelectromagnetic Inflation,
Phys. Lett. B **642**, 302 (2006)

[Mez04]   M. Mézard,
Theory of Random Solid States,
in *Stealing the Gold: A Celebration of the Pioneering ...*
P. M. Goldbart, N. Goldenfeld and D. Sherrington eds.,
Clarendon Press, Oxford (2004)

[MH09]   M. J. Mortonson and W. Hu,
Evidence for Horizon-Scale Power from CMB Polarization,
Phys. Rev. D **80**, 27301 (2009)

[MM05]   J. Martin and M. Musso,
Stochastic Quintessence,
Phys. Rev. D **71**, 63514 (2005)

[MM06a]   J. Martin and M. Musso,
Solving Stochastic Inflation for Arbitrary Potentials,
Phys. Rev. D **73**, 43516 (2006)

[MM06b]   J. Martin and M. Musso,
Reliability of the Langevin Perturbative Solution ...
Phys. Rev. D **73**, 43517 (2006)

[MMR04]   S. Matarrese, M. A. Musso and A. Riotto,
Influence of Super-Horizon Scales on Cosmological Observables ...
JCAP **405**, 8 (2004)

[MOL89]   S. Matarrese, A. Ortolan and F. Lucchin,
Inflation in the Scaling Limit,
Phys. Rev. D **40**, 290 (1989)

[MP91]   M. Mézard and G. Parisi,
Replica Field Theory for Random Manifolds,
J. Phys. I **1**, 809 (1991)

[MPV87]   M. Mézard, G. Parisi and M. A. Virasoro,
Spin Glass Theory and Beyond,
World Scientific Lecture Notes in Physics Vol. **9** (1987)

[MW06]   S. P. Miao and R. P. Woodard,
Leading Log Solution for Inflationary Yukawa Theory,
Phys. Rev. D **74**, 44019 (2006)

[Nat97]   T. Nattermann,
Theory of the Random Field Ising Model,
cond-mat/**9705295** (1997)

[NDH+09]   M. Nolta, J. Dunkley, R. S. Hill, et al.
Five-Year Wilkinson Microwave Anisotropy Probe Observations: ...
ApJ. **180**, 296 (2009)

[NNS88]   K. i. Nakao, Y. Nambu and M. Sasaki,
Stochastic Dynamics of New Inflation,
Prog. Theor. Phys. **80**, 1041 (1988)

[OW02]   V. K. Onemli and R. P. Woodard,
Super-Acceleration from Massless, Minimally-Coupled ϕ^4,
Class. Quant. Grav. **19**, 4607 (2002)

[Pol87]   A. A. M. Polyakov,
Gauge Fields and Strings,
Harwood Academic Publishers (1987)

[PP95]   J. V. Porto III and J. M. Parpia,
Superfluid ^{3}He in Aerogel,
Phys. Rev. Lett. **74**, 4667 (1995)

[PPR78]   R. A. Pelcovits, E. Pytte and J. Rudnik,
Spin-Glass and Ferromagnetic Behaviour ...
Phys. Rev. Lett. **40**, 476 (1978)

[PS79]   G. Parisi and N. Sourlas,
Random Magnetic Fields, Supersymmetry, ...
Phys. Rev. Lett. **43**, 744 (1979)

[PS95]   M. E. Peskin and D. V. Schroeder,
Introduction to Quantum Field Theory,
Westview Press (1995)

[Rey87]   S.-J. Rey,
Dynamics of Inflationary Phase Transition,
Nucl. Phys. B **284**, 706 (1987)

[Rio02]	A. Riotto, *Inflation and the Theory of Cosmological Perturbations*, hep-ph/**210162** (2002)
[Riz02]	T. Rizzo, *Ultrametricity between States at Different Temperatures* ... cond-mat/**207071** (2002)
[Ros04]	R. Rosenfelder, *Pfadintegrale in der Quantenphysik*, lecture notes, ETH Zürich, corrected and extended version (2004)
[RTV86]	R. Rammal, G. Toulouse and M. A. Virasoro, *Ultrametricity for Physicists*, Rev. Mod. Phys. **58**, 765 (1986)
[SK75]	D. Sherrington and S. Kirkpatrick, *Solvable Model of a Spin-Glass*, Phys. Rev. Lett. **35**, 1792 (1975)
[SMI06]	Y. Sakamoto, H. Mukaida and C. Itoi, *Stability of Fixed Points in the $(4+\varepsilon)$-Dimensional* ... Phys. Rev. B **74**, 64402 (2006)
[SO92]	D. J. Sellmyer and M. J. O'Shea, in *Recent Progress in Random Magnets*, ed. D. Ryan, World Scientific, Singapore (1992)
[SS85]	M. Schwartz and A. Soffer, *Exact Inequality for Random Systems: Application* ... Phys. Rev. Lett. **55**, 2499 (1985)
[Sta80]	A. A. Starobinsky, *A new Type of Isotropic Cosmological Models without Singularity*, Phys. Lett. B **91**, 99 (1980)
[Sta82]	A. A. Starobinsky, *Dynamics of Phase Transition in the New Inflationary Universe* ... Phys. Lett. B **117**, 175 (1982)
[Sta86]	A. A. Starobinsky, in *Field Theory, Quantum Gravity and Strings*, Vol. **246** of *Lecture Notes in Physics*, Springer (1986)

[SVP+03]	D. N. Spergel, L. Verde, H. V. Peiris, E. Komatsu, M. R. Nolta, ... *First Year Wilkinson Microwave Anisotropy Probe* (WMAP) ... ApJ. Suppl. **148**, 175 (2003)
[SW67]	R. K. Sachs and A. M. Wolfe, *Perturbations of a Cosmological Model and Angular Variations* ... ApJ. **73**, 147 (1967)
[SY94]	A. A. Starobinsky and J. Yokoyama, *Equilibrium State of a Self-Interacting Scalar Field* ... Phys. Rev. D **50**, 6357 (1994)
[Tho30]	L. H. Thomas, *The Radiation Field in a Fluid in Motion*, Quart. J. Math. **1**, 239 (1930)
[Tis42]	L. Tisza, *Supersonic Absorption and Stokes' Viscosity Relation*, Phys. Rev. **61**, 531 (1942)
[TW05]	N. C. Tsamis and R. P. Woodard, *Stochastic Quantum Gravitational Inflation*, Nucl. Phys. B **724**, 295 (2005)
[van81]	N. G. van Kämpen, *Itô versus Stratonovich*, J. Stat. Phys. **1**, 175 (1981)
[Vil82]	J. Villain, *Commensurate-Incommensurate Transition with Frozen Impurities*, J. Physique Lett. (France) **43**, 808 (1982)
[Vil83]	A. Vilenkin, *Birth of Inflationary Universes*, Phys. Rev. D **27**, 2848 (1983)
[Wei71]	S. Weinberg, *Entropy Generation and the Survival of Protogalaxies* ... ApJ. **168**, 175 (1971)

Bibliography

[Wei72]   S. Weinberg,
Gravitation and Cosmology: Principles and Applications ...
1st eds., Wiley-VCH (1972)

[Wie92]   K. J. Wiese,
Die Zeta-Funktion für ein su.-sy. nichtlineares Sigma-Modell,
Diploma thesis, Universität Heidelberg (1992)

[Wie02]   K. J. Wiese,
Ungeordnete Systeme,
lecture notes, Universität GHS Essen (2002)

[Wie05]   K. J. Wiese,
Why one Needs a Functional Renormalization Group to Survive ...
Pramana **64**, 817 (2005)

[Woo05]   R. P. Woodard,
A Leading Logarithm Approximation for Inflationary ...
Nucl. Phys. Proc. Suppl. **148**, 108 (2005)

[WV00]   S. Winitzki and A. Vilenkin,
Effective Noise in a Stochastic Description of Inflation,
Phys. Rev. D **61**, 84008 (2000)

[WW90]   E. T. Wittaker and G. N. Watson,
A Course in Modern Analysis,
4th eds., Cambridge University Press, p. 95 (1990)

[YF50]   C. N. Yang and D. Feldman,
The S-Matrix in the Heisenberg Representation,
Phys. Rev. D **79**, 972 (1950)

[YGH+94]   U. Yaron, P. L. Gammel, D. A. Huse, et al.
Neutron Diffraction Studies of Flowing and Pinned Magnetic ...
Phys. Rev. Lett. **73**, 2748 (1994)

[Zin05]   J. Zinn-Justin,
Quantum Field Theory and Critical Phenomena,
Oxford Science Publications, Forth Edition (2005)

[ZLF98]   C. Zeng, P. L. Leath and D. S. Fisher,
Absence of Two-Dimensional Bragg Glasses,
cond-mat/**9807281** (1998)

i want morebooks!

Buy your books fast and straightforward online - at one of world's fastest growing online book stores! Environmentally sound due to Print-on-Demand technologies.

Buy your books online at
www.get-morebooks.com

Kaufen Sie Ihre Bücher schnell und unkompliziert online – auf einer der am schnellsten wachsenden Buchhandelsplattformen weltweit! Dank Print-On-Demand umwelt- und ressourcenschonend produziert.

Bücher schneller online kaufen
www.morebooks.de

VDM Verlagsservicegesellschaft mbH
Heinrich-Böcking-Str. 6-8   Telefon: +49 681 3720 174   info@vdm-vsg.de
D - 66121 Saarbrücken   Telefax: +49 681 3720 1749   www.vdm-vsg.de

Printed by Books on Demand GmbH, Norderstedt / Germany